C and the 8051 Programming for Multitasking

Thomas W. Schultz

Purdue University
West Lafayette, Indiana

P T R Prentice Hall
Englewood Cliffs, New Jersey 07632

Library of Congress Cataloging-in-Publication Data

Schultz, Thomas W.
   C and the 8051 : programming for multitasking / Thomas W. Schultz.
     p.   cm.
   Includes bibliographical references and index.
   ISBN 0–13–753815–4
    1. C (Computer program language)   2. INTEL 8051 (Computer)–
 –Programming. I. Title.
   QA76.73.C15S368   1993   92–5117
   005.265—dc20   CIP

Editorial/production supervision
  and interior design: *Harriet Tellem*
Cover design: *Wanda Lubelska*
Prepress buyer: *Mary E. McCartney*
Manufacturing buyer: *Susan Brunke*
Acquisitions editor: *Karen Gettman*
Editorial assistant: *Barbara Alfieri*

© 1993 by P T R Prentice Hall
Prentice-Hall, Inc.
A Paramount Communications Company
Englewood Cliffs, New Jersey 07632

The publisher offers discounts on this book when ordered
in bulk quantities. For more information, write:

  Special Sales/Professional Marketing
  Prentice-Hall, Inc.
  Professional Technical Reference Division
  Englewood Cliffs, New Jersey 07632

Printed in the United States of America
10 9 8 7 6 5 4

ISBN 0-13-753815-4

Prentice-Hall International (UK) Limited, *London*
Prentice-Hall of Australia Pty. Limited, *Sydney*
Prentice-Hall Canada Inc., *Toronto*
Prentice-Hall Hispanoamericana, S.A., *Mexico*
Prentice-Hall of India Private Limited, *New Delhi*
Prentice-Hall of Japan, Inc., *Tokyo*
Simon & Schuster Asia Pte. Ltd., *Singapore*
Editora Prentice-Hall do Brasil, Ltda., *Rio de Janeiro*

The Lord will command the blessing on you . . .
in all that you undertake.

Deuteronomy 28:8

Contents

Chapter 5   **Routines**   **90**

Chapter 6   **Modular Program Development**   **116**

Chapter 7   **8051 Family Hardware**   **156**

Chapter 8   **Real-Time Ideas**   **182**

Preface

The advanced development tools of today are a dramatic shift from the primitive techniques used for the first microprocessors. The growth of software packages and the demand for increasing size and power in the 16- and 32-bit personal computers have fed each other. Perhaps the love of code efficiency and simplicity has been lost—particularly for general-purpose computing. However, there still remains a corner of the field where *small* dedicated microcontrollers are best. While small microcontrollers run more slowly, access much less memory, and seldom have elaborate math capability, they happen to be *much less expensive*! This book is not meant to take away from the high-power developments; rather, it is intended to illustrate the ideas and techniques that set the *embedded* control field apart from the large-computer-system emphasis that draws so much attention.

While the area of *microprocessor* software has seen a shift from assembly language to high-level languages, simulators, and emulators, very little has been published regarding practical applications of *multitasking* to control projects. Most programming books stop before that point. But larger projects, which now often involve several programmers, could benefit greatly from both the modular programming and multitasking techniques discussed here.

For almost 20 years I have worked with dedicated microprocessor projects, and for almost 15 years I have been teaching in that area. About eight years ago, I started teaching (and have continued to develop) a senior-level course in Electrical Engineering Technology on real-time multitasking. It began with RMX 86, and over the last six years the course has shifted to DCX and BITBUS (built around the 8044 member of the 8051 family of microcontrollers). From the start, few people, particularly in academic circles, seemed interested in PL/M and 8051 assembly language. But, with the decline of the 8085 and 6800 families, academic users have cast about for a new processor. At the same time, some of the EET department's courses shifted to the 8051 and the C language. This book started to look possible. At the same time, the cries of my students for a "book to go to" finally got to me. A book discussing multitasking in more specific detail is also under development.

I am particularly grateful to my wife, who has constantly encouraged me. She also unfailingly took me to task over grammatical errors and helped remove them. In addition, I wish to express thanks to Siegfried Bleher of Franklin Software for his overwhelming encouragement and help. Thanks are also due to Gary Muhonen of Iota Systems and Charles Behrmann of CMX for input on the last few chapters. Charles Wales graciously allowed the inclusion of material on guided design. Finally, John Poindexter allowed his article on 8051 boards to be reproduced and Charles Larson agreed to refine his article on C compilers for this book.

Introduction

The two objectives of this book are (1) to show you how to program for embedded applications, and (2) to help you develop the mindset of multitasking. The focus is on using the C language. This book concentrates on the sort of applications where the attached hardware is more likely to be switches and solenoids than disk drives and keyboards. The computer may have only one job to do all its life. Often called a *real-time* system, it must be ready at just the moment it is needed and is usually handling a number of inputs and outputs at the same time.

WHY THE 8051?

This book is focused on the 8051 family microcontroller. Although the original 8051 has been on the market for about 10 years, the family (mostly in its CMOS versions) is still a frequent choice for middle-of-the-road projects needing embedded control. If you face *high-end* (high-speed) applications, there seem to be new 16- and 32-bit controllers coming out yearly with no clear "winner" yet, and for *low-end* (high-volume, low-speed) applications, there is still a place for 4-bit controllers such as National's COPS series. But in between fits the 8-bit controllers, which have a high level of flexibility without the high cost. The leaders in this middle area are currently the 68HC11 and the 8051. The latter is so common in low-to-medium-performance applications that the processor core is even available in ASIC (Application Specific Integrated Circuit) design libraries much the same way as a flip-flop or counter design would be added to a custom integrated circuit. The 8051 has become a very real *standard* in the microcontroller world. In addition, the 8051 family now includes a number of high-performance-on-one-chip versions described in Chapter 7. Along with the hardware changes, the software tools have improved with new C compilers and multitasking operating systems.

If you are a practicing *designer* or *system developer* you will find this book's hands-on treatment, full of specific examples, provides the starting point for more elaborate designs. This book can "get you up to speed" in the C language, help you switch over to the 8051 for some new design, help you start using multitasking, or just provide ideas for driver-level interfacing for embedded hardware. By seeing trade-offs that go into efficient programs, your own code will become more efficient. If you are already experienced in other languages and familiar with microcontroller hardware, learning a language through working examples such as this book provides should be much quicker than through a formal teaching text. Many of the examples may be directly adapted to new projects.

Additionally, if you are a *student* in a microcomputer *application* course or are facing a design project, this book will give you practical programs using the 8051 with the sorts of hardware commonly used in embedded systems. There are programs to input from switches and output to lights, or to input from a 4 × 4 keypad and output to a low-cost LCD display module. Other examples use stepper motors, speech chips, and A-D and D-A converters. There are even a few schematics of 8051 family systems you can copy in Chapter 7.

PLANNING

Perhaps less dramatic than a new language or a new processor, but in the long run more important, is project planning. You can really learn planning only by experience—reading about it is not enough in itself. If you take "learning to plan" too seriously, it can look formidable, but this book eases the process with a set of practical suggestions on *how to get started*. Good designers are always looking for new ideas!

Chapter 1 illustrates the general approach to **project planning** by taking a hypothetical project from start to finish. In addition to suggestions about software development, it gives an overview of typical hardware development steps for embedded systems and briefly describes tools such as simulators and emulators. It also includes the easily overlooked system integration phases of a project!

PROGRAMMING LANGUAGES

We cannot discuss real applications without actual programs. Although this book could serve to teach programming from ground zero, I assume you are *already* familiar with *some* assembly language or *some* high-level language. If you have never been exposed to *any* programming, then I suggest you start with a more elementary text.

What language should be used? Three languages are commonly used with the 8051 family hardware—C, PL/M, and Assembly. For the 8051 family, other than the language manuals themselves, assembly language, PL/M, and C have not been discussed very much anywhere. Only recently have books discussing 8051 assembly language begun to appear. All the books on C seem to deal with running data processing examples, which are inappropriate to the 8051. As far as feasible, the programming examples in this book are provided in *all three* languages. This may be particularly helpful if you are following the wholesale migration to C.

Even if knowing assembly language is not your goal, seeing and understanding a little of it will help you understand the machine-specific limitations of the 8051 that affect efficiency in any language. For example, knowing the assembly language instructions makes the advantage of on-chip RAM for variables quite obvious, because you can see that accessing any off-chip variable requires several instructions to set up the accumulator and data pointer. All the examples in Chapters 2–6 have been compiled or assembled and either tested with a simulator or run with actual hardware, so you can be reasonably confident that they work. Incidentally, the multitasking examples of Chapters 9–11 are *derived* from working code and have been reviewed by the operating system vendors, but they have not been tested—be careful!

PROGRAMMING

Chapters 2 through 5 cover C, Assembly, and PL/M for the 8051 family. It will help to have some previous programming experience. If you are only moving over to the 8051 and keeping the same language, you may want to look at Appendix B. To keep the examples simple for those just beginning, the first six chapters do not use any I/O or hardware features of the 8051 except ordinary parallel ports. Mixed in through these chapters, however, are suggestions to improve your programming, which may help even experienced programmers.

Chapter 2 introduces the three languages and illustrates the **logical and arithmetic** operations that are important to embedded applications. Most microcomputers for general use have a "flat" memory space—code, data, tables, and so on, all having different addresses—so it is important that you understand the different kinds of variables and the different types of memory space in the 8051. The precedence of operators is also discussed.

Chapter 3 reviews the **branching and looping** constructs, which are essential to any structured programming approach. The idea of *structured* programming is explained and such points as the difference between a loop test at the start or end of a loop are included.

Chapter 4 gets into **pointers** or **based variables,** which are fundamental to general-purpose subroutines or functions (and probably were stuck in at the end of your programming course). This chapter also goes into structures and arrays. Examples include the use of look-up tables with interpolation compared to direct calculation.

Chapter 5 covers **functions, subroutines, and procedures**—the pieces that contribute to modular, understandable programs. This chapter also goes into passing values to and from routines.

MODULAR PROGRAMMING

Chapter 6 goes into scope of variables, using multiple files in developing software, and the mixing of languages. Seldom are more than a few pages devoted to these topics in programming texts, yet they are the key to modern programming. When several programmers work together to develop a project, the modular approach is no longer just a technique for switching to assembly when a high-level language becomes too slow; it is the *key* to well-organized programs that can be easily maintained.

8051 HARDWARE

Following the language sections, Chapter 7 pauses to explain some of the very specific hardware common to most 8051 family members. Although not a substitute for the manufacturer's data sheets, this chapter introduces the special function registers (SFRs), which control the timers, interrupts, and serial ports. Because the vast majority of applications add at least off-chip EPROM, several examples of small and expanded systems are given with a discussion of some of the trade-offs involved. A brief description of the variety of additional hardware features within the 8051 chip family is included. Appendix C includes a description of several commercially available 8051 boards and systems.

MULTITASKING

If you are going to write efficient embedded applications code, you must venture forth on *new way of thinking* about programs. Probably your earlier programming courses concentrated on single tasks; if they were data-processing-related there were probably no real-time requirements involved. Chapter 8 introduces the ideas surrounding having one processor seem to do several things simultaneously. Switching to a *task* mentality can improve your whole approach to complex projects.

Chapter 9 shows the use of a timer (and interrupt) to produce a scheduler and a pulse generator. This is done with "scratch" programming and then goes on to do the same things with commercial operating systems (DCX, DCE, RTX,

USX, CMX, Byte-BOS and RTXC). Chapters 9–11 repeat virtually the same example in each of the systems so you can see both the similarities and the differences among them.

Chapter 10 discusses communication between tasks on the same controller. In addition to the simplest methods using flags and shared variables, this chapter describes signaling, message passing, and resource management in the various multitasking operating systems.

Chapter 11 picks up on interrupts. In addition to the priority and context-switching issues, it illustrates ways to use flags to avoid long interrupt routines. Then it shows how the commercial multitasking systems handle external interrupts.

Chapter 12 describes multiple processor and distributed control systems. In addition to methods involving the standard serial port, BITBUS, DCE-COMM, and FILBUS are described.

Chapter 13 is a brief review of multitasking with a discussion of the costs and benefits of real-time operating systems.

Appendix A gives an overview of assembly language instructions for the 8051.

Appendix B describes some of the software details of the 8051. It includes a discussion of compilers for C written by Charles Larson. There is also a section on language-switching hints—8085 versus 8051 assembly; PL/M 80 versus PL/M 51; ANSI standard C versus C with 8051 extensions.

Appendix C describes some of the hardware associated with 8051 development. It includes a reprint of an article by Richard Poindexter surveying commercial 8051 boards.

Appendix D lists all the known vendors of 8051-related products with current addresses and phone numbers as well as brief comments or descriptions of products.

REVIEW AND BEYOND

1. What are the 3 programming languages of this book?
2. Why is the 8051 an appropriate controller for study?
3. From your own experience, give several examples of embedded controllers in consumer products.

Chapter 1

Project Planning

DO YOU HAVE A PROJECT?

Before getting into the details of programming of embedded systems, it is important to consider the planning and development process. Although it would be possible to speak of planning in broad generalities, I prefer to use a recently completed *specific* project as a model of the steps of planning, which ought to go into *any* microcontroller project. The actual application has been obscured enough so I can talk frankly without reflecting on the companies involved, but the planning process has been left intact. I hope the decision steps will help you and line up with your own experiences.

Projects seldom seem to come fully defined. Usually the *customer*, whether from outside or within your own company, thinks the project will be quite easy. *After all, it is not much different from what is currently being done. Even if it involves significant changes, it just requires a few of those little chips; don't you just wire them together, and that does it? Oh yes, software; I have a brother-in-law who writes software for an insurance company, and he should be able to get around any problems.*

On the other hand, *you* may have discovered an application just waiting for a microcontroller, but the current approach is working and there is no big push to change. The way the application is being done is OK, but it *could* be done much more *cheaply*, more *flexibly*, and more *reliably*. In such a situation, be very careful to not push in with too many promises until you understand the current approach quite thoroughly. Even then, don't confuse parts cost with total system cost. There may have been *years* of experience with the present approach, which shows up any shortcomings, and *no* experience with your proposed alternative.

For an example, picture a small company that wants to get into the microcontroller business. Because of the nature of their business, they have all sorts of sales contacts with users of such control equipment. Occasionally there are applications where the customer's present controller is too expensive, unreliable, or just not suitable for new products coming along. This small company has opportunities that most new ventures see only in their dreams. *But* they have no experience in microcontroller design and electronic development in general. They need some help.

For a specific project, imagine a company that manufactures envelope handling systems for mass mailings like that shown in Fig. 1–1. In the system, as the envelopes come out of the box, they are addressed by a printer and then picked up by a handler. They are then moved down a conveyor to where the flap is opened

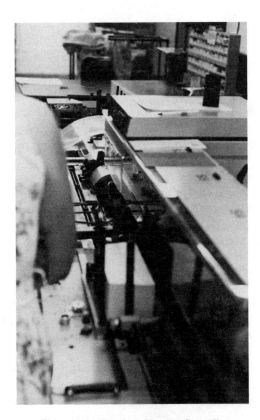

Figure 1–1   Envelope-Handler Controller

up, a letter or set of material inserted, and the envelope sealed. In the existing system, the finished envelopes are run out on a tray for manual zip-code sorting and bundling. The system works fairly well most of the time, but there are times when noise causes the controller to reset, and the system has to be manually cleared and restarted. The controller is a custom unit supplied by another company, which has been (supposedly) not too responsive to the manufacturer's complaints and doesn't seem to want to spend more development money. A small company getting a foot in the door supplying controllers to this manufacturer would have a ready-made market.

SYSTEM PLANNING

One of the classic errors of planning is failing to **define** what is to be accomplished. You just *know* what is required and immediately start to do it. Only after much effort, time (and money) are expended do you slow down enough to see that you are developing a great solution to a *different* problem. It is much better if you first sit down and determine exactly what is (and is not) desired by the *customer*. In many cases, the customer may not know what is difficult or easy to accomplish. Negotiation *at the start* can determine if a difficult feature is actually necessary, as well as determining whether to include alternative features the customer may not have considered. In any case, a solid agreement at the start can avoid the **creeping features** problem as the project progresses and the customer asks if you can include this or modify the goal to do this.

There is an entire decision-making/problem-solving process that is formalized in various management texts and classes. One version of the process is outlined in Fig. 1–2.

The critical first step, often done in group settings, is to *identify the problem*. It is highly wasteful to develop a detailed solution to the *wrong* problem. The specific goal of the decision work is to outline a plan for the small company mentioned above to develop a product which will meet the controller needs of the envelope-handler manufacturer better than the present controller (and thus provide an entry into the custom controller market for the small company).

Not enough is known to proceed, so an **information-gathering** visit to the manufacturer is undertaken. The handler is impressive to watch with envelopes zipping along and being picked up to have letters inserted. The setup process for the machine is demonstrated and proves to be quite complicated. The current controller has a host of individually labeled membrane pushbuttons with a complex set of multiple functions defined by the current mode of the system. The button functions are particularly illogical when setup is going on, but there *is* the benefit of a single, specifically labeled button for each function during normal operation. It is very much a *custom* controller. Coupled with that, it is packaged in a plastic box. The circuit boards (there are several) are packed in with a wrap of aluminum foil, suggesting some efforts to add shielding. Careful examination shows that the

THE DECISION-MAKING PROCESS*

Identify the Problem The problem must be separated from its symptoms and properly identified (a critical step). Then you should be able to

State the Goal of Your Work The goal defines the specific result you expect to accomplish through your decision-making work. The next step should be to

Gather Information This step may also occur at other places in the process. The questions *Who?*, *What?*, *When?*, *Where?*, *How?*, and *Why?* should be useful. The answers are separated into columns of *What is?* and *What is not?* The next step is a

Component Analysis This is used to identify the major components of the problem/situation or system. The key question here is *"What major factors can be changed?"* Then you should

Generate Possible Solutions These will achieve the goal. The question here is *"How can each component be changed?"* When this creative, divergent task is completed, it is appropriate to specify the

Constraints and Assumptions Constraints limit what can be done. Assumptions help simplify the problem. The next step is to make a

Choice A choice is made by applying the constraints and any other criteria that are relevant to the possible solutions and carefully examining the positive and negative consequences of different options. A table is sometimes used to help organize this work and the solutions may be ranked from best to poorest. The best solution is selected for further work. This is also a critical step in the decision-making process because a great deal of detailed work will follow. This work begins with an

Analysis of the chosen solution. An excellent way to perform this step is to identify all the components that must be considered and ask as many questions as you can about the choice using the words *Who, What, When, Where, How and Why.* After all the components have been identified, it is time to begin the

Synthesis In step 1 the analysis questions are answered; in step 2 a detailed solution is devised. When this work is completed, an

Evaluation is performed to see if the plan satisfies the goal and meets any criteria that are appropriate. Here again the positive and negative consequences should be examined. If the evaluation shows that the solution is not acceptable, it may be necessary to return to the possible solution step, make another choice, and proceed again through the analysis, synthesis, and evaluation steps. If the work is acceptable, a plan to evaluate the implemented solution should be devised. Then

Recommendations may be made and a

Report prepared. If the work is approved, you will

Implement the plan and prepare to

Check the Results.

*Reprinted courtesy of Center for Guided Design (Charles E. Wales), 137 Engineering, West Virginia University, Morgantown, WV 26506-6101

Figure 1–2   Decision-Making Steps

triac drivers for the pneumatic solenoids, which run the indexing equipment, are right in with the microcomputer chips. Inductive spikes are a real possibility and will be conducted very close to the microcomputer itself. There is opto-isolation of the triacs themselves, which provides some degree of protection. With an added power-line filter at the input as well, it is obvious that some effort has been made to improve the existing unit.

Having gathered information, the next step is to determine the *"what, when, where, why, how"* answers that go with journalism and also help in problem definition. The approach is to say, *"What is* the problem?" and, *"What is not* the problem?" Then, *"When is* it a problem?" and *"When is it not* a problem?" A variety of answers come out of this, but it shows that the problem is not with everything. For example, in the envelope controller, the problem is not with the basic functioning of the controller. Many end users have had the envelope systems for years without major complaints. Some of the isolation techniques are good. The price to the OEM is not considered too high—actually it is much less than that of a commercial programmable controller such as a company like Allen Bradley might sell. The custom button array of the controller package is considered an absolute essential because it gives the OEM a prominent name on the system. Finally, we conclude that the main problem relates to the lack of noise immunity, which causes occasional failures. Coupled with that is the lack of friendliness for setup and the lack of flexibility.

The step of **component analysis** consists of determining which major factors can be changed. With the envelope handler, because there is a large installed base of envelope sorters, the manufacturer is not very interested in radical changes, but *claims* to be open to whatever might be proposed. Because the existing controller supplier is not at all open to licensing or changing their design, it is clear that a new design is in order for this start-up company.

Next we **generate possible solutions**. The control application, although it looks impressive, actually involves less than two dozen control outputs and a few opto-sensor inputs to determine if the envelopes have gotten hung up. Speed is not at all a problem. Because the entire process is effectively just a series of timers, we consider using the 8051 family microcontroller because we have experience with it. Obviously other controllers could be used instead, but the base of previous experience suggests that the 8051 family is about as inexpensive as any, it is multiple-sourced, in wide usage, and likely to remain on the market a long time. It has built-in timers and interrupts, which should reduce the chip count, and it is expandable to 64Kbytes of code space if necessary. Actually what we are doing is shortcutting some of the "possible solution" process in the choice of a microcontroller!

In embedded controller applications, it is important to determine what aspects of the project have the **most uncertainty**. *Those* parts need the most attention *early* in the project to be sure they can be achieved. Perhaps a keyboard is to be used with which no one is familiar. Perhaps a math algorithm is planned that may run too slowly. The focus of the early activity should be on whatever is most uncertain.

HARDWARE/SOFTWARE TRADE-OFFS

The decision process outlines the steps for **constraints and assumptions** and the **choice**. A critical part of this for embedded control projects is the hardware/software trade-offs. In general, if software can do it, the cost will be less. Software, once developed, can be used over and over at no added cost, but hardware has to be purchased and assembled for each unit. If the system is fairly busy, however, the choice of more software gets into processor-speed and memory-space questions.

One of the first steps in getting specific is the decision of where to be hardware- or software-intensive. Will all the switches be debounced with flip-flops? Or will the software be written to do a periodic scan of the switches and decide the state has changed only after two successive readings agree? The former is easier for the programmer, but the latter does not involve the *repeated* cost of the flip-flops and increased board real estate. The software-intensive approach may involve more program memory, but memory has gotten cheaper—EPROM smaller than 8Kbytes is rare and may be no less expensive than much larger ones. Besides, unused space in an EPROM does not bring a refund from the chip supplier!

Other trade-offs relate to the user interface to the unit. Since LCD modules have become so inexpensive, the use of large arrays of custom buttons has been declining. A menu approach is much more flexible because any message desired can be sent to the display with a software routine. A change of user language is easily done for international markets. Some displays even have versions supporting a Japanese character set. Software-defined buttons could allow flexibility— put the buttons just below a two-line display, and the lower line can show the current function and name of each button. Incidentally, such an approach makes the hardware much more universal because it can be adapted to other customers with only a software change.

In the envelope-handler example, the decision to use an 8051 microcontroller probably comes prematurely! No one is *totally* objective. The justification is that the start-up company has *no* experience with any controller chip. Because all their projects are likely to be relatively small ones with simple on/off control, it seems unlikely they will soon have projects needing regular computers or high-performance controllers. The two or three other applications they are considering also could be easily done with an 8-bit controller. If they were to get into image processing, digital filtering, etc., they would have to hire an entire new department rather than upgrade their present people! Because the 8051 has built-in timers and UART, it can reduce the chip count from the earlier envelope controller design. Cost is not critical, but there is no good reason to purposely choose a higher performance approach.

The question of noise immunity is a more difficult one because it is not known precisely how the disturbances enter. Because the routing of the solenoid driver printed circuit traces is close to the microcomputer, it is assumed that electromagnetic radiation is the problem rather than power-line transmitted noise. Two approaches are chosen. First, put the unit in a steel box to improve the magnetic shielding, and arrange the package to have a middle wall separating the low-level signals from the AC side. Ideally the opto-isolators would be right at the wall between the two halves. Much more experimentation is necessary to really determine the cause, and the time schedule does not allow that luxury.

The second prong of the noise "war" is the careful use of cold and warm start routines. Some sort of watchdog timer arrangement will recognize when the processor is behaving strangely and force a warm restart. Ideally the software can then recover from a random disturbance. Usually noise corrupts an instruction as it is being read into the processor and causes the processor instruction fetches to get out of step. Once the instruction count is off by one, the next "instruction" may actually have been the second byte of the previous instruction. From there it is a sort of roulette game as to whether the program sequence will recover or go jumping off to places unknown. If the program has been periodically saving the state of the process (what is going on), when the watchdog timer fails to be reset on time because the program is lost, the timer can cause a jump to a restart location (warm start), which assumes the process is still in the last stored state. There is particular safety where the state is stored in an EEPROM because its very slowness to take new values means it can't quickly be corrupted by a lost program.

Another common hardware/software trade-off relates to analog/digital interfacing. It is straightforward to simply connect up an A-D chip to a port or use an 8051 family member with a built-in converter. But an alternative is to bring the analog voltage into a V-F (Voltage to Frequency) converter. Because a frequency or period can be measured by the internal clock/timer, the voltage can be digitized with only a single input pin and a small V-F chip. Because the accuracy of the reading is determined by the amount of time spent counting, the 12-bit accuracy of the V-F chip can be realized, but a 6-bit reading can be obtained quickly. The V-F converter is quite noise-immune because the frequency over time is related to the *integral* of the incoming voltage—noise pulses just make small changes in the time the next pulse of the V-F comes along.

Incidentally, the terms "analog" and "digital" date back to the early days of electronics. An analog computer was a collection of operational amplifiers, integrators, and differentiators that used the voltage to represent the analogue of physical parameters such as the position of a car above a road with springs, mass and shock absorbers represented by capacitors, resistors, and inductors. The importance of the specific voltage value is the key to analog electronics. Digital computers (probably related to fingers as digits) ignore the exact value and represent a "1" as, say, any voltage above 2.4 volts and a "0" as any voltage below 1.0.

Finally, a common trade-off relates to the question of keeping track of time with an external clock/calendar chip. By using an external chip, it is easy to keep track of the time of day, month, year, leap year, and so on, but the cost is an extra chip and crystal as well as possible address decoder and buffer. In the example, the calendar features are not significant and the internal timer of the 8051 is more than sufficient. In projects where the low standby power of a separate chip is unimportant, to still use separate clock/calendar chips to avoid the relatively minor programming required to track time using the internal 8051 timers seems a poor choice. If the application already involves a real-time interrupt for other timing functions (as in Chapter 9), then the calendar function is truly a minor software item.

HARDWARE PLANNING

Having defined the envelope-handler *problem*, it seems that the project needs buttons for user input, ports and optoisolated triacs for solenoid drivers, and some sort of display. A timer function is needed to keep things in sequence, and a flexible data storage structure is necessary to sort out the user setup inputs into sequence commands. All of these hardware decisions rely on a good knowledge of electronic devices on the market. Such knowledge comes by experience but is not the focus of this book. Although real hardware will be used throughout this book in program examples, it is the *programming* and *multitasking* that are the main concerns. For the envelope handler, once the overall hardware is determined and that design job is proceeding, serious software planning can start.

SOFTWARE PLANNING

There is an approach to programming called **top-down**, which relates to the idea of understandability and the use of routines. Essentially it is an approach that starts with the broad overview (the "top") and then moves ("down") to the details. Using such an approach, you *first* write the main program (the "main" function in C). In doing so, you make liberal use of routines (discussed in Chapter 5) that you haven't written yet, and assume that these routines will perform in specified ways. For example, if you need a printing device, call your (at that moment nonexistent) **printf** routine with a pointer to the desired message. If you need an A-D reading, you assume some yet-to-be-written routine will exist to return the reading when called.

Once you've written the main program, the "down" of "top-down" consists of writing the first level of routines. These in turn may use other routines, which you will write later. Finally, you write the detailed drivers—the most basic routines.

Such an approach looks good on paper, but you will find most hardware-oriented projects have details requiring practical experience. At the beginning, it may be better to use a **bottom-up** approach to software development, which starts with the most basic drivers. *First*, write a routine to initialize the display and show a single character. *Then* write a canned message program. *Then* write routines to accept varying messages. *Then* write fixed and scrolling message routines. *Then* perhaps write messages with constant text and variable values intermixed. *Finally,* go on with the rest of the project!

The best solution may be an **ends-in** approach that maintains an overall perspective while developing the most basic details. That avoids the *"Where do I go from here?"* sense that comes from the bottom-up approach, as well as the *"You mean* *the hardware can't do it?"* problem of the top-down approach. From any direction, there is no substitute for experience in this phase of project development.

RULES FOR TASK PARTITIONING

In planning software, it is strongly suggested that you *divide the overall job into tasks.* The term **task** simply means *the thing to be done* and corresponds to the various routines that go with modular programming. Sending a message to a display or getting an interpolated value from a look-up table is an example of an individual task. If several programmers are involved, task partitioning may be used to even up the programming load while letting individuals work where they can be most productive. In partitioning tasks, it is easiest to go bottom-up and first set up **driver tasks** for the individual hardware components. If a joystick is to be used, you might make an interrupt task to interface the joystick. The task could set a flag bit when changes are observed, or a separate background routine could be written that returns the most recent readings. A display task could be passed the actual information or brief codes that refer to a library of canned messages. The partitioning may even have a display driving task, which is called only by a code-to-message conversion task. At some point the question of how general purpose to make the driver tasks will need to be resolved.

Once the driver tasks are defined, then the **"thinking" tasks** can be planned. If there is significant math processing to do, the math might be made a background task. If serial communication is required, then a single task to interface the "outside world" may be in order. The general idea is to break the overall job into logical, isolated pieces.

Finally, there may be **management tasks** which define the sequence of operations or determine when user input is calling for some other mode of operation. These tasks will probably be isolated from the I/O hardware and interact with preprocessed information in the form of flags or simple bytes of information.

When defining tasks, it may be appropriate to establish priorities. Chapters 8 and 10 go into a little of this. In all the planning it is important to include testing and system integration so that the pieces come together smoothly.

For the envelope sorter, it is decided that the LCD display interface will be two separate tasks—one task will be the initialization of the display and a second task will take the pointer to a message and convert it to a sequence of LCD drive commands and data. A third task will scan the user-input keys. At least initially, there are a large number of keys, so scanning will be broken up into pieces to distribute the load among the interrupt cycles. If one quarter of the keys are scanned every 10mSec, then the time to the next scan of the same key will be 40mSec. That is long enough for a key to stop bouncing without missing any button pushes or appearing to have delay. Several tasks are defined to get the different user commands onto the display and take the resulting entries into a data table. Another task will take the key scan results and convert them into coded values. Then one more task will convert the user setup entries into actual sequential operation times, because the user will set the duration of pneumatic activations, but the actual times when they are turned on and off must be measured from the start of the cycle. Then, as shown in the first example of Chapter 9, the main cycle-timing (scheduler) task is set up. It will run off the timer interrupt every .01Sec because that is the smallest significant time increment. Every .01Sec the scheduler checks to see if any outputs are due to change (and changes them) as well as doing one quarter of the button scan.

REAL-TIME DATA FLOW DIAGRAMS

When multiple tasks are involved in a job, the tasks are often interrelated, and it is useful to show this in a "block diagram" manner. In its simplest form, the conventional task diagram includes rounded rectangles for the tasks, with arrows showing the information that passes between tasks, and rectangles for the actual hardware interfaced by the tasks. The diagram of Fig. 1–3 shows the planned tasks for the envelope handler.

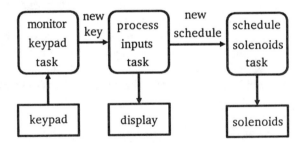

Figure 1–3   Envelope-Handler Task Diagram

DEFINING TASK INTERFACES

Once you've outlined and named the tasks, it is important to define *carefully* the **interfaces between tasks.** Chapters 8 through 11 discuss this further. Where several people are writing program modules, it is particularly important that they agree about the uses of flags or variables. *What does each flag mean? How long should a task wait for a flag? Is it necessary to use flags to determine if the shared data has been picked up? Will things go right during development even if some tasks are not yet written?*

    The task interfaces for the envelope-handler example involve the data to be shared with the display task as well as the way to determine if the button state has changed. The actual sequence times are shared by the background user-interface tasks and the interrupt-driven solenoid-output task. When a value is changed by the user, as background processor time becomes available, the values will be sequentially updated. If the updating caught up with the operation, the cycle parameters will be changed throughout the rest of the current cycle and then into later cycles. The only danger with updating during operation is that the change in "on" and "off" times might cause an *off* to be missed in the current cycle. This relates to the idea of *protected regions or resource management*, which is also part of multitasking. For this case, the problem scan be avoided either by requiring the operation to be stopped before new parameters can take effect, or by using two lists and setting up the new one while the system continues to use the old one.

HOW BUSY IS THE PROCESSOR?

There are excellent tools that can evaluate time in loops, etc., but these simulators and emulators will not be discussed in any detail. Several vendors are listed in Appendix D if you want more information. When using the multitasking methods of Chapters 8 through 12, if processor overloading is suspected, a suggestion would be to devise a temporary background task to replace the regular background task. The new task could simply count free time. The faster the free-time-count increases (perhaps you could even run a stepper motor off the overflow to make it more graphic!), the more free time there is. If the counting stops, then there is no free time, and the system is obviously **overloaded**.

If too many things need to be done on a continuing basis, the processor will never be able to catch up. Major math algorithms can be sent to a math chip, or faster controllers are available. It is also possible to go to several processors with a serial communication scheme between them; this is discussed in Chapter 12. In the end, the real test of the system is, "Is it fast enough?" Where people are the sources of inputs to the controller or the users of outputs, "fast enough" is a subjective question, but it still needs to enter into the planning.

In the project involving the envelopes, there does not appear to be any problem with overload because the running of the system is given higher priority than the display update or user interface processing. The scheduler interval of 10mSec should be slow enough to easily get through the preprocessed list of start/stop times. It might have trouble if the entire user-input-to-actual-times conversion had to be completed during one interrupt. The same concern is behind the division of the scan into four pieces, but is probably unnecessary. Probably the only measures of *fast enough* for the example are (1) the speed of the display update when the user pushes buttons, and (2) the requirement that all the outputs change on time. The speed of detecting an envelope jam is probably not critical because the system only needs to respond in the approximately one second index time.

SYSTEM INTEGRATION

Planning the way in which the pieces will be gradually developed and smoothly fit together is an often-neglected part of planning. *How will the individual driver tasks be tested when nothing else is completed? How can a supervisory task be tested when none of the worker tasks are ready yet? How can the basic interaction be tested before adding all the bells and whistles?* These are all questions of system integration.

Probably the best way to start is to **test the hardware** before testing any software. Write a short test program to read and write to the ports. See which logic level turns on the LEDs, which pins come out where on the cables, what sequence is necessary to initialize the LCD, and so on. Despite all your presumably careful hardware design and schematic layout, you need to be sure the hardware works the way you supposed it would.

Having done that, **start small**. Write a *simple* program to read in from the keypad and store in a variable location that can be sent out to a parallel port. Make a task to put a *simple* message on the LCD. Get the speech chip to say a short phrase over and over.

In the example project, the integration steps involve the following:

1. Get the breadboard to power up without any chips, and check the presence of correct voltages at the proper pins.
2. Install the processor (with power off while installing!), and check for a clock signal or ALE.
3. Burn and try a simple program that reads a port and echoes to another port.
4. Write and debug a program to send a fixed message to the display.
5. Write a program that echoes the keypushes to the display.
6. Get the real-time interrupt routine going, and have it send a count to the display.
7. Develop the sequencing routines, and test the triac (solenoid driver) outputs (with neon bulbs).
8. Work out the conversion routines for the user input/output sequence.
9. Take the finished unit to the manufacturer to test on an actual machine.
10. Debug, revise, revise, package, etc.

TESTING AND MAINTAINING?!

One of the easiest planning steps to overlook is the planning of testing and maintenance. In the hardware design stage, it is easy to forget that troubleshooters will need a place to connect to ground and power for using a simple logic probe or scope. If a logic analyzer may be used, it is good to have a row of pins accessible for connecting to the various buses. Certainly, at least the processor and EPROM should be socketed. The mechanical design of the packaging should include ways to easily extend out the boards for access while the controller is running. Circuit layouts seldom seem to be made with any logical order to help in following the traces, so a good component layout diagram and schematic are essential. This could include silk-screened component layout on the board and separate diagrams of component placement and orientation. There are much more elaborate testability schemes including ones involving a special mode on the buffer chips or additional "glue logic" devices, but this is beyond the scope of this book. It bears watching for serious high-volume designs, however.

SELF-TEST ROUTINES

A well-planned project will include self-test and self-calibrate features. The software could start by going through its memory space looking for locations that are bad. The entire EPROM space could be evaluated for a checksum to determine if some location might have become corrupted. But, beyond the usual computing-hardware sorts of tests, special test modes could be built into the program so that, say, the D-A converter can be made to output a staircase waveform, the pushbuttons can be individually shown on the display when pushed, each of the outputs can be activated directly by a push of a button, and so on. Such test routines would make it easier to diagnose the bad components (given that the processor works). The degree of effort placed here will depend on the expected production quantities (if it is to be the only one, it may be easier to troubleshoot it in the lab), the anticipated repair philosophy (if it is a throwaway, why bother), the location of the repair shop, and the skill level of the individuals charged with maintenance.

CLOSURE ON DECISION MAKING

The process above, in an embedded-control setting, includes the remaining steps of the decision-making sequence—**analysis, synthesis, evaluation,** and **recommendations.** For the envelope handler it was done over several meetings and involved the comparison of the existing controller with the proposed design. A prototype was developed and shown to the customer for feedback. At the time of this writing, some 15 months after the start of the project, the company has switched to the new controller and shipped several dozen envelope-handler systems with the new controller installed. In the for-what-its-worth category, the LCD display was replaced with a similar module using a light-emitting technology, several iterations of the custom membrane keyboard were required, and the software has grown to push the 64K limit on code due to a high degree of custom setup features added (creeping features!) and is in need of a serious overhaul to reduce its size. It would be a prime candidate for the multitasking systems of Chapters 8–12 if there were time and resources to start over.

REVIEW AND BEYOND

1. What are some of the hazards of neglecting planning?
2. What do you think are advantages of doing problem definition in a group rather than alone?
3. What arguments can you make for doing a new controller design when a current controller is doing an adequate job?
4. Why is a V-F an attractive alternative to an A-D in some applications? When would this not be the case?
5. Why is it easier to develop software with a *task* perspective?
6. List at least three hardware features to build in that can help with maintenance and troubleshooting.

Chapter 2

Three Languages

WHY THESE THREE?

You *can* learn programming from this book, but because the fundamentals are covered rather quickly, you should have at least a passing acquaintance with some assembly language or some high-level language. The goal here is to demonstrate programming for practical *embedded* applications. With the 8051 microcontroller family, I know of support for only four languages—assembly, PL/M, C, and BASIC. BASIC will be discussed only briefly. The other three languages are the subject of the detailed discussion from here through Chapter 5. Chapter 6 then goes into the process of getting a source code transformed to an actual program running on the 8051.

C is the language originally used to write the UNIX operating system and until recently it remained closely tied to UNIX. It is a *structured language*, which will produce compact code. The structure is marked out with braces { } rather than words and the language makes use of special symbols seldom used in normal writing. There can be detailed control of many machine-level functions without resorting to assembly language. C, however, can be *so* condensed that someone else may have to study a program for some time to figure out what is going on.

Only with careful attention can you write programs that nonprogrammers can understand. Beginning about 1985, C has been available for 8051 microcontrollers, and there are at least six different companies selling 8051 C compilers. Not all these compilers produce efficient code oriented toward taking advantage of the 8051. Where the C examples that follow become compiler specific, it is Franklin's version 3 compiler that is in view. Some details of the other C compilers are included in Appendix D.

Assembly language for the 8051 is much like other assembly languages. The instruction set is a bit more "powerful" than that of the first generation microprocessors, but the different memory regions inherent in the 8051 family complicate things significantly. The move, math, logical, and branching instructions are generally similar to those of most other processors. Although it is an annoyance to learn another assembly language, the process is not difficult if you are already proficient in one.

PL/M is a proprietary Intel language that has been available for their processors beginning with the 8080. It is much like PASCAL, but with roots in PL-1. Like C, it is a *structured* language, but it uses very readable arrangements of key words to define the structure. The PL/M compiler produces fairly compact code much like a good assembly language program. C and PL/M can be much easier to use than assembly language, because the compilers and linker/locater manage the details of variable locations and moving between memory regions. PL/M is a "high-level *assembly* language" in both a negative and positive sense. There can be very detailed control of the code generated, but, for the 8051 family, PL/M has no support for complex mathematics, floating-point variables, or trigonometric functions.

WHY NOT BASIC?

Before proceeding, let me explain why BASIC *doesn't* fit. **BASIC** is routinely available on PCs and is a common first programming language. It is for just that purpose—a *basic* introduction to programming. There are few "getting started" details involving variables. The use of a new variable name defines it as a variable for the rest of the program. It is very easy to use. Errors can stand out as you complete each line, rather than showing up when the program is finished. Also, it is easy to make changes line by line. There are three definite reasons why interpreted BASIC *may* not be a good choice for embedded applications.

First, by its **interpreted** nature it is slow—each line must be converted to machine code as it comes up when running. The process of getting from English to machine code is a complex one and takes lots of computing time. There are **compiled** versions of BASIC (QUICKBASIC) which, if you will, do the interpreting ahead of time, but there are none known for the 8051 family.

Second, in simplifying the use of variables, all variables are usually made full floating-point values. Adding 2+2 is a full floating-point math operation. This is a complex set of instructions, so running time gets long. Even with more elaborate compiled BASIC, if floating-point math must be supported, resulting programs will be slow and large.

Third, no *compilers* for BASIC on the 8051 are known to the author. Only one version of the language *interpreter* for the 8051 family is known; it comes preprogrammed in an 8052. Although BASIC is quite suitable for applications where the ease of programming is more important than efficiency or speed, it will not be discussed further in this book.

VARIABLES

Because computing usually deals with numbers, the question arises, *"How should numbers be represented?"* The choice of variable type or data type is far more critical with the 8051 than with math-oriented computers. Only the first two types in Fig. 2–1 are directly supported by the machine instructions (and assembly language). Regardless of the high-level language used, though an operation may look quite simple in your program line, the manipulation of the more complex variables will be handled by *series* of machine instructions. Using floating-point variables in particular will add considerably to computing time and program size. When such precision is *necessary*, the fact that C automatically includes the library routines is convenient, but a poor programmer may use a larger variable type unnecessarily. In assembly language the overhead of such an error is obvious because *you* must write the routines to handle the variables; it is immediately clear that the code becomes more complicated. PL/M 51 has little automatic overhead because it adds support for only the unsigned integer (16-bit word) variable. But a C program can easily pull in large libraries to handle the large variable types. Later, at linking time, the programmer discovers the program is slow running and perhaps too large to fit in the available code space.

One variable type is the **bit**. A bit is either 1 ("true") or 0 ("false"). Bit variables that take advantage of the 8051 hardware must be in the on-chip RAM. Although assembly and PL/M 51 support the bit hardware directly, in C the use of those bits are an extension to the standard (portable) language. The bit *fields* within an integer for ANSI standard C are quite a different thing and are very implementation dependent.

Data Type	Size	Range
bit	1 bit	0 or 1
unsigned char		
(PL/M "BYTE")	1 byte	0 to 255
unsigned int		
(PL/M "WORD")	2 byte	0 to 65535
(signed) char	1 byte	-128 to + 127
(signed) (int)	2 byte	-32768 to +32767
(signed) long	4 byte	-2147483647 to + . .
unsigned long	4 byte	0 to 4294967295
float	4 byte	6 digits e-37 e+37
double	8 byte	10 digits

Figure 2–1 Data Types for Variables

COMMENT LINES: Probably the most elusive error for PL/M and C programs is the failure to end a comment line properly. The start with / * will not end until a * / is encountered. If the * or / gets left off, the compiler will faithfully regard the following lines as comments until it reaches the (correct) end of the *next* comment. Usually the error comes near the start and misses some of the variable definitions, producing strange errors all through the program. The best way to detect the problem is to look at the *list* file (such as in Fig. 6-3 or Fig. 6-4) and look for program lines where the indentation depth numbers are missing. Probably that is because they are treated as comments.

SIGNED AND UNSIGNED: If you use both types you may pull in two forms of the library routines, which will use up more code space. If speed is important and negative numbers are not involved, make everything unsigned.

USE UNSIGNED CHAR: Use unsigned char (BYTE) variables whenever possible since they are directly handled by the hardware. Use bit variables for the same reason. *Signed* char variables take only one byte but manipulation will pull in code to test the sign.

C is not inherently very bit oriented—there is not even a direct binary notation for constants. Instead, a hexadecimal notation is usually used. Most of the *8051* compilers add some way to define and use the bit-addressable features of the 8051 family, but technically that makes the language not fully portable—you can't directly address bits on an 8086, for example. Ultimately, you must decide how machine-independent your code needs to be. When the 8051 C compilers are most machine-*in*dependent they produce the largest, most inefficient code!

char/BYTE variables are 8-bit values, which fit ideally in the 8051 because it handles 8 bits at a time. They have values from 0 to 255 (**unsigned**) unless they are **signed** (not supported in PL/M 51). Then the most significant bit, **msb**, is the sign. A 1 represents a negative, so signed and unsigned representations are the same for 0 to 127. Negative numbers are represented by 2's complement notation, which makes a −1 to be a 11111111 in binary and a −2 to be 11111110. Interestingly enough, that is just what would happen if a binary counter was at 0 and you subtracted 1 or 2 from it. The notation was chosen to be quite hardware-friendly. When you want to multiply or divide, then the sign issue is more complex, and the C compiler might bring *libraries* into play.

WORD/int variables are 16-bit values. Unlike the 8080 and 8086 families, they are stored with the *most significant* byte at the *lower* address. *Signed* values again have the msb as the sign bit and use 2's complement notation. Several of the machine instructions make the multi-byte addition and subtraction quite straightforward. Similar to the int variables are the 4-byte (32-bit) **long** variables.

More complicated exponential representations are the **float** and **double**. They have both sign and magnitude for the exponent and the mantissa. Any math operations with them involve library functions with varying degrees of efficiency depending on the compiler. Indeed, this seems to be one field where the "compiler advertising war" is fought most intensely. When most of the control applications are quite hardware intensive, this may not be the important area! The implementation of floating-point storage is not specified by the ANSI C standard, so the arrangement of the sign and exponent parts could vary among compilers.

SHORTHAND—#DEFINE AND LITERALLY

Many programmers choose to develop shortcuts to save typing. These are easily done with **#define** or **LITERALLY** expressions at the top of the listing. After a few examples using the official words, the examples will switch to several shorthand abbreviations to shorten the listings. Specifically, in C, u c h a r will be used for uns i gned char and u i nt for uns i gned. Likewise, in PL/M AUX will be used for AUXILIARY. The definition lines will sometimes be shown in the examples, but you should realize that the body of the example will look nonstandard (and more compact!).

MEMORY SPACES

In discussing the 8051 family of microcontrollers, the types of memory space need to be mentioned also. Unlike large computers, there are at least three memory locations that can have the same address.

First there is the **code** space, which is where program instruction code and other unchanging information goes. The finished program probably would go into an

EPROM. In the running of the program there is no writing *to* such space because the program should not modify itself. Code (or, in PL/M, CONSTANT) space is the logical place to keep canned messages used for interaction with a user/operator.

The second type of memory space is the internal **data** (or, in PL/M, the MAIN) RAM. It is between 64 and 256 bytes long depending on the particular processor (the original was 128), and is part of the microcontroller chip. By today's standards that is not very much memory, but a variety of machine instructions can access it quickly. Somehow at least one of the expanded 8051 family members described in Chapter 7 has 512 bytes of internal RAM. The "external" memory can physically be part of the chip, but those values must always be moved into data space before they can be used. The *internal* data memory is a good place to "park" variables temporarily for computations, as well as a place to keep frequently used variables.

Finally, there is external **xdata** (or in PL/M "AUXILIARY") memory that is not on the 8051 chip itself. There can commonly be 8 or 32K bytes in one added chip—up to 64K bytes is directly addressable. The machine instructions to access this memory must bring the value into internal memory before putting it to use—a process involving several machine instructions in itself—and then return the result to external memory. External memory is the place for less frequently used variables and for collected data waiting to be processed or forwarded to another computer.

ON-CHIP RAM: Use on-chip RAM for frequently used variables. In PL/M they go there ("MAIN") by default. In C you can either use a SMALL memory model which puts all variables there by default or, depending on the compiler, force them there by the use of idata or data keywords. In assembly you simply avoid the movx instruction and the xseg designations.

Some C compilers support two seldom used memory types which PL/M doesn't support but which are available in assembly language— **idata** and **pdata**. The former is indirectly addressable internal memory (above 127). The latter is xdata with a one byte address where port 2 is being carefully preserved for I/O. Both are of no interest for single-chip applications and most expanded-memory applications because it is difficult to even *buy* a 256-byte memory chip. It *could* apply to an application that only added off-chip I/O devices in that memory space.

The allocation of memory space is different for each of the three languages, as shown in Fig. 2–2. All the examples through Chapter 6 have been tested with Intel's assembler (ASM51), Intel's PL/M compiler (PL/M 51), or Franklin's C compiler (C51). Many aspects of memory space in programming as well as the details of using assemblers and compilers are covered in more detail in Chapter 6.

```
1   #define PORTA XBYTE[0xffc0];
2   bit flag1;
3   code char table1[]= 1,2,3,"help",0xff;
4   idata unsigned int temp1;
```

[1]Some machine-specific additions to C show up here since the language was not originally designed with different memory spaces in mind. The **#define** for PORTA is a special function specific to Franklin C. Other 8051 C compilers also allow direct access, but the code words may be different. This defines a *fixed* location (probably an I/O port) addressed as external RAM.

[2]This defines a single bit variable (which can only be in internal RAM).

[3]This defines a group of 8 bytes (in ROM) that begin at the address to be later assigned to TABLE1.

[4]A 2-byte space for storage in internal RAM is set up here.

Figure 2–2a   Memory Allocation—C

```
    mybits segment bit
    myrom segment code
    myiram segment data
    rseg mybits
1   flag1: dbit 1
    rseg myrom
2   table1: db 1,2,3,'help',0ffh
    rseg myirom
3   temp1: ds 2
4   xseg at 0ffc0h
    porta: ds 1
5   end
```

[1]While it is possible to reserve a memory location in an absolute sense, the choice of location may, instead, be left for a *linker/locater program*, as discussed in Chapter 6. The segment designation determines the *type* of address space used. The first portion in the example defines a single bit variable (which can only be in internal RAM).

[2]Here a group of 8 bytes (in ROM) that begin at the address to be later assigned to TABLE1 is defined.

[3]A 2-byte space for storage in internal RAM is set up here.

[4]This defines a *fixed* location (probably an I/O port) addressed as external RAM.

[5]Any assembly language source program *always* concludes with the directive **end**.

Figure 2–2b   Memory Allocation—ASM51

SINGLE AND DOUBLE QUOTES: In C it is easy to confuse them. Single quotes do not add the end of message character (usually a zero) to the character string and can only be up to four in length (a long variable).

21

```
1   TEST:DO;
2    DECLARE FLAG1 BIT;
3    DECLARE TABLE1 (*)
      BYTE CONSTANT(1,2,3,'help',0FFH);
4    DECLARE TEMP1 WORD MAIN;
5    DECLARE PORTA BYTE AT(0FFC0H) AUXILIARY;
6    DECLARE P1 BYTE AT(090H) MAIN;
     END TEST;
```

[1]Any PL/M program begins with a module name (here TEST) and a DO. With PL/M, the variable **declaration** alerts the compiler to use the appropriate set of machine instructions to address the particular memory spaces desired. Here a bit is defined first. Each block (started with a DO or a PROCEDURE) concludes with an END. This is the most basic part of a *structured* language.

[2]The first portion in the example defines a single bit variable (which can only be in internal RAM).

[3]This "DECLARE" sets up a group of 8 bytes (in ROM due to "CONSTANT") that begin at the address to be later assigned to TABLE1.

[4]A 2-byte space for storage in internal RAM is set up here.

[5]This declares a *fixed* location (probably an I/O port) addressed as external RAM.

[6]This declares one of the internal ports. The address is fixed in the hardware of the chip.

Figure 2–2c   Memory Allocation—PL/M

PORTS

To have realistic embedded examples, it is simplest to start with parallel I/O ports. The 8051 has only memory-mapped ports, meaning that there are no specific I/O instructions—the ports are just memory locations, which happen to show up as signals on external pins. There are four standard 8-bit I/O ports in the basic 8051 members, but often 2 1/2 of them are used up in expanding the EPROM or RAM space. This, as well as the addition of I/O ports off-chip, is discussed in Chapter 7. Figure 2–2 shows how the software sets up on- and off-chip ports. It takes very special treatment to convince the C compiler to put a variable at a fixed location. Both C and PL/M have include files that define the internal special function registers of the 8051 members.

PORT USAGE: The internal ports of the 8051 family are bit-addressable, while most external ports are only byte-addressable. In assigning ports to hardware, where there is a choice, put individual control lines and single I/O signals on internal ports and byte-wide I/O on external ports. That will greatly simplify the addressing and avoid a lot of the logic involved in masking bits.

EXAMPLE: SWITCHES TO LIGHTS

I start with a somewhat complicated example of Figs. 2–3a through 2–3i that shows a program that could work on the setup of Fig. 2–3a. Chapter 6 goes into details of program compiling and linking, and Chapter 7 discusses some of the hardware design issues. As envisioned, this program reads in from 8 switches connected to (on-chip) P1, stores the readings in a 10-byte array, shows the most recent reading on 8 light-emitting diodes (LEDs) connected to P3, and waits 1/10 second to repeat the process. Discussion of the millisecond time delay is found in Chapter 3 (Fig. 3–9).

If you are curious about *how* the high-level languages perform the operations, for this example the resultant assembly code is shown in Figs. 2–3j and 2–3k. The code option can be requested for any C or PL/M compilation, but you would usually include it only when very specific details are in question. For example, in this case, it was used to discover that the reversal of the order of assignment to P3 and ARRAY(I) had no effect on code efficiency. For these listings,

DEEP INDENTATIONS: It is generally unwise to use indentations over 4 spaces because the sucessively deeper indentations for nested blocks will leave the comments to fold around on the list file if you print on an 80-character (8 1/2" wide) printer. It is suggested that you use a depth of 2 as is shown in most of the examples.

TABS: There can often be a problem with tab characters in a source file where the editor and the printer have different tab stop settings. Reconfiguring printers is often messy or out of your control. Rather than that, consider replacing tabs with spaces (which are universally understood). Some editors have a feature that will substitute spaces whenever you hit a tab as well as features that will follow a carriage return with enough spaces to bring the start of the next line under the start of the previous one. That is very handy for nested indentation (which is a must for any good structured program).

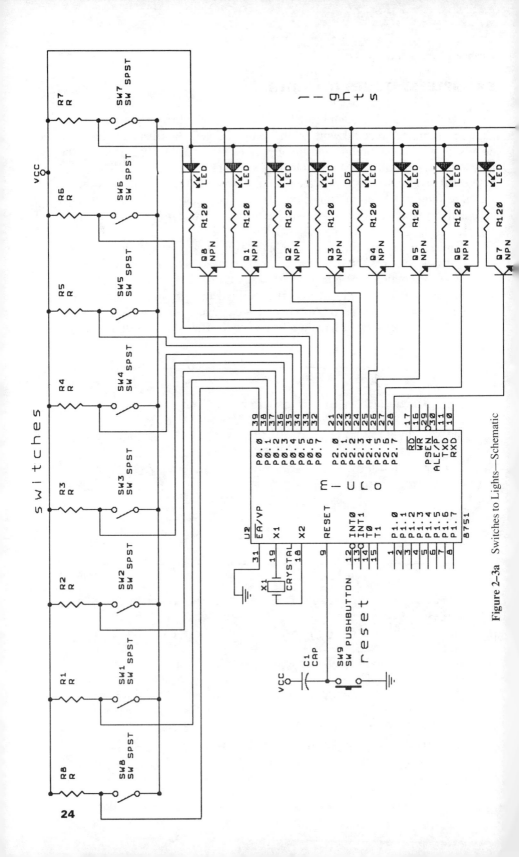

Figure 2-3a Switches to Lights—Schematic

24

Figure 2–3b Switches to Lights—Picture

lts-switches

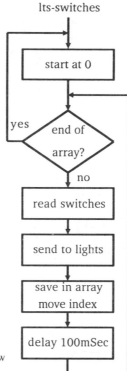

Figure 2–3c Switches to Lights—Flow Chart

```
1   #include <reg51.h>
2   void msec(unsigned int);

    void main()
    {
      unsigned char array[10];
      unsigned char i;
3     while (1)
      {
4       for (i=0; i<=9; i++)
        {
5         array[i]=P3=P1;
2         msec(100);
        }
      }
    }
```

Figure 2–3d   Echo Switches to Lights—C-1

the actual machine code values have been stripped off with an editor but would usually be to the left of the assembly listing. Franklin C and PL/M put the code listing separate from the high-level source, whereas some other C compilers put the resulting assembly instructions interleaved with the C instructions.

```
1   #include <reg51.h>
2   void msec(unsigned int);

    void main(){
      unsigned char array[10];
      unsigned char i;
3     while (1){
4       for (i=0; i<=9; i++){
5         array[i]=P3=P1;
2         msec(100);
        }
      }
    }
```

Figure 2–3e   Echo Switches to Lights—C-2

```
1   #include <reg51.h>
2   void msec(unsigned int);

    void main(){
      unsigned char array[10];
      unsigned char i;
3     while (1){
4       for (i=0; i<=9; i++){
5         array[i]=P3=P1;
2         msec(100);}}}
```

[1]With C there are a variety of commonly accepted ways to arrange the braces ({ }) as well as a variety of indentations. The separation into lines and the indentation are not important to the compiler, but lines and indentation can help the *readability* greatly. The first version is quite correct and is preferred by some who emphasize the structure of the language, but it becomes difficult with the deep indentation to fit lines on 80-column pages when the nesting becomes deep. The style of the third version is quite compact but raises complaints among some programmers, so the second style will be used throughout this book. For long blocks the various braces by themselves can be confusing to type at the right nesting depth. If you carefully observe the shallow (two space) indentation, the structure of the third form is easily visible. Obviously, if you totally ignore indentation depth, *any* system will be difficult to follow!

The include file (or header file) is necessary to define the two port names used in this example. It actually defines all the registers and ports internal to the 8051, and other header files define the registers for other members of the family described in Chapter 7.

[2]The m s e c delay routine is **prototyped** here, meaning the actual function is found in some other module (see Chapter 6) and this line simply tells the compiler the details needed here to call the routine. Chapter 3 goes into the details of functions including m s e c s, but for now consider this just a way of getting a delay between readings.

[3]With C, an endless loop can be either w h i l e (1) or f o r (; ;). For unknown reasons, one reference says the latter is the preferred method. For understandability, either one is usually defined as "forever."

[4]The f o r block shows that i is to start at 0 (the initialization, which here sets i to point to the first element of the array). The program is to keep on looping as long as i <= 9 (the test condition—"i s less than or equal to nine."). Looping is discussed in Chapter 3. At the end of the loop, i is to be incremented by one each time (the reinitialization part). C does not restrict the reinitialization to incrementing and allows entire expressions involving other variables in any of the three parts.

[5]The two equal (assignments) work because the = evaluates from right to left. That is to say, the value of P 1 is assigned to P 3 *before* the value of P 3 is put into the array.

Figure 2–3f   Echo Switches to Lights—C-3

BITWISE OPERATORS

Control applications more often use bitwise operations than arithmetic. With external input and output ports, it is often desirable to read or change one bit of a byte without affecting the other bits. Perhaps this one bit turns a motor on or off while the other bits of the port may activate warning lights or tell an A-D con-

```
7
1   extrn code (msec)
2   mycode segment code
2   mydata segment data

2   rseg mydata
      array: ds 10

2   rseg mycode                 6
3     start:mov r0,#array ;set array pointer
      again:mov acc,P1
        mov P3,acc
        mov @r0,acc ;store in array
        mov r2,#0 ;high byte
        mov r1,#100 ;low byte
4       lcall msec
        inc r0 ;point to next loc.
5       cjne r0,#array+10,again ;end?
        sjmp start ;reset to start
      end
```

Figure 2–3g   Echo Switches to Lights—ASM51-1

verter to start a conversion. Particularly with multitasking, it may be impossible to predict in advance what the *other* bits of the port will be. As already mentioned, some ports are bit-addressable (those directly on the 8051 family chips, for example), but most other add-on ports respond only as entire bytes. This is where bitwise operators come into play. Figure 2–4 lists the bitwise operators for the three languages. Incidentally, C makes a sharp distinction between bitwise operators and logical operators that PL/M determines from the context. Logical operators are discussed later in this chapter.

Assume in the examples of Fig. 2–5 that PORTA is an external byte-addressable port needing the third bit from the bottom (bit 2, because counting

---

CONFUSING BITWISE AND LOGICAL PRECEDENCE: When a program waits for a bit to change (a button has been pushed), the **masking** may be done wrong: DO WHILE PORTA AND 08H > 0; or while (PORTA & 0x8 > 0) would seem to stop when the fourth bit goes low. But since the > has precedence over the AND/&, the right side evaluates as true (1 or FF), and there is really no testing for the intended bit.

7

```
1  extrn code (msec)
2  mycode segment code
2  mydata segment data

2  rseg mydata
       array: ds 10

2  rseg mycode      6
3      start:    mov r0,#array          ;set array pointer
       again:    mov acc,P1
                 mov P3,acc
                 mov @r0,acc            ;store in array
                 mov r2,#0              ;high byte
                 mov r1,#100            ;low byte
4                lcall msec
                 inc r0                 ;point to next loc.
5                cjne r0,#array+10,again ;end?
                 sjmp start            ;reset to start
       end
```

[1]The details of msec are discussed near the end of Chapter 3 and the details of combining program pieces from separate files is the subject of Chapter 6. For now it is included for completeness and can be thought of as "magic" to get a delay between readings.

[2]The segment lines define what type of memory space is to be used in linking the rseg (relocatable segments). Here the array will be located in internal RAM and will take up 10 bytes. The mycode segment is to be put in code space (EPROM). As written, after being assembled, this code and data will need to be **located**, which is the preferred method when writing with multiple program modules. It is possible to include an org (originate) instruction to direct the assembler to fill in specific jump addresses and put together an *absolute* piece of code. Such code would automatically "own" that space, and the link/locate utility program would have to use other space for code from **relocatable** modules. With multiple modules to combine in multitasking applications, it can be an annoyance keeping several absolute modules from overlapping. Chapter 6 discusses modular programming in more detail.

[3]The first mov (move) instruction puts the starting address of array into the internal register r0. As r0 increments, it will point to successive locations in the array.

[4]The time delay is not defined here and might involve a simple delay loop or the internal timers of the 8051 family. Details of the latter are discussed in Chapters 7 and 9.

[5]When r0 reaches array+10, the cjne (compare/jump if not equal) instruction will bring program flow back to start. There r0 will be reset to point to the first byte of array again.

[6]The question of style for assembly is open to debate. Some would argue that this indentation is too shallow for the labels. Others would argue that the comments could be slid over to the end of the mnemonics. The assembler doesn't care.

[7]Two different styles for assembly programming are presented. It is a matter of personal taste as to how deep one indents and whether one keeps labels, mnemonics, and comments vertically aligned. In order to save space, the tighter style will be used in this book.

Figure 2–3h Echo Switches to Lights—ASM51-2

```
   TEST2:DO;
1  $INCLUDE (reg51.dcl)
   DECLARE ARRAY (10) BYTE MAIN;
   DECLARE (I,J) BYTE MAIN;

2  MSEC:PROCEDURE (X) EXTERNAL;
    DECLARE X WORD MAIN;
   END MSEC;

3  DO WHILE 1;  /*FOREVER*/
    DO I=0 TO 9;
4     ARRAY(I),P3=P1;
2     CALL MSEC(100);
    END;
5   END;
   END TEST2;
```

[1]This include file is necessary to let the compiler know about the registers and ports in the 8051. Specifically, here it identifies P1 and P3.

[2]Again the details of external program pieces is the focus of Chapter 6, and the details of the delay routine are covered in Chapter 3. For now, consider it a way of getting a time delay between readings.

[3]The DO WHILE 1 line gives an endless loop since a test of 1 for a 1 in the least significant bit will always be true. The condition to leave the loop will never happen (actually the compiler is "smart" enough not to go through the motions and produces a simple jump instead). The looping constructs are discussed in Chapter 3.

[4]The output of the reading and the storage can be done in one line as shown. The comma has a distinctly different use than with C. If this was not done on one line, the port would be read twice or the reading should be put into a temporary storage location.

[5]A structured language would not generally use a jump or a goto command since the DO/END block takes care of that.

Figure 2–3i   Echo Switches to Lights—PL/M

starts with bit 0) set high and needing bit 6 set low *without affecting any other bits.*

ROTATE AND SHIFT

In addition to the operations already mentioned, there are two bit-related operators that rearrange a byte. First is the **rotate**. If you think of a byte as a collection of bits from the left (msb—most significant bit) to right (lsb—least significant bit) order, you can correctly see that a rotate right moves all the bits to the right one. That is to say that each bit moves over to a less significant

```
?C0001:
        CLR     A
R       MOV     i,A
?C0003:
R       MOV     A,i
        SETB    C
        SUBB    A,#09H
        JNC     ?C0001
        MOV     R7,P1
        MOV     P3,R7
R       MOV     A,#array
R       ADD     A,i
        MOV     R0,A
        MOV     @R0,AR7
        MOV     R7,#064H
        MOV     R6,#00H
E       LCALL   _msec
R       INC     I
        SJMP    ?C0003
```

Figure 2–3j   Resulting Assembly Code—C

```
WHILE?1:
F       MOV     I,#00H
  DO?3:
F       MOV     A,I
        SETB    C
        SUBB    A,#09H
        JNC     DOEND?4
        MOV     A,P1
        MOV     P3,A
        MOV     R6,A
F       MOV     A,I
F       ADD     A,#ARRAY
        MOV     R0,A
        MOV     @R0,AR6
F       MOV     X+0001H,#64H
F       MOV     X,#00H
F       LCALL   MSEC
F       MOV     R0,#I
        MOV     A,#01H
        ADD     A,@R0
        MOV     @R0,A
        JNC     DO?3
  DOEND?4:
        SJMP    WHILE?1
```

Figure 2–3k   Resulting Assembly Code—PL/M

Logical Operation	Assembly Instruction	PL/M Operator	C Operator
NOT	cpl a	NOT	~
AND	anl a,#	AND	&
OR	orl a,#	OR	|
EXCLUSIVE OR	xrl a,#	XOR	^

Figure 2–4   Logical/Bitwise Operators

```
1   extern xdata unsigned char PORTA;

    void main(void){
      PORTA=(PORTA & 0xbf) | 0x04;
    }
                    2      3
```

[1]This line declares the existence of a variable named PORTA which is defined elsewhere. A separate module is necessary which would set the port at a specific address. In this case it is intended to have an assembly language module handle such details, as will be discussed in Chapter 6. If you would avoid this, there is an alternate approach shown in Fig. 2–8 which allows direct addressing of ports without linking in another module.

[2]Hexadecimal constants start with 0x. There is no notation for binary numbers, so they must be expressed in hex or octal (be aware that, without the x, a leading 0 on an integer constant means octal!). It is possible, in defining a constant, to force a definition of "long" or "floating" by appending a suffix of l or f.

[3]The & and l associate from left to right. The parenthesis is unnecessary, but it is always safer to include it.

Figure 2–5a   Bit Set and Clear—C

```
    PORTA equ 0ffc0h
1   cseg at 2000h
      mov dptr,#PORTA
      movx a,@dptr        ;get the present reading
      anl a,#10111111b    ;set bit 6 low
      orl a,#00000100b    ;set bit 2 high
      movx @dptr,a        ;output new values
      end
```

[1]For variety, this example uses the cseg directive rather than naming the module and then making it relocatable, as was shown in Fig. 2–2c. The difference relates to issues at linking time and is discussed in Chapter 6.

Figure 2–5b   Bit Set and Clear—ASM 51

```
TEST2:DO;                        1
      DECLARE PORTA BYTE AT(0FFC0H) AUXILIARY;
      PORTA=(PORTA AND 0BFH) OR 4H;
END TEST2;            2
```

[1]Hexadecimal (base 16—0123456789ABCDEF) notation for constants must start with 0-9 and end with H. Binary notation must end with B and have only 1's and 0's. The declare line identifies the address FFC0 as the storage location for a BYTE variable named PORTA. All external ports for the 8051 family are memory-mapped and there are no I/O instructions as would be with the 8080/85/86.

[2]The AND and OR associate from left to right. The parenthesis is unnecessary, but it is always safer to include it.

Figure 2–5c   Bit Set and Clear—PL/M

place. The result is a divide by 2 in the same way that moving a decimal point for a base ten system is a divide by ten. A rotate left would be a multiply by 2 in the same way.

What happens to the last bit in the row? Assembly language has two kinds of rotates. The first, using r r or r l, is a simple 8-bit rotate where the top bit goes around to the bottom or vice versa. A second assembly rotate, r r c or r l c, includes the carry bit. If the carry is cleared, then a 0 rotates *into* the byte. If multi-

FAST MULTIPLY AND DIVIDE: Depending on the cleverness of the compiler, a shift can be faster than a multiply or divide. Some compilers will actually substitute directly for a multiply or divide by 2, so it may not save anything. It will be somewhat obscure to a novice reader, too. However, when you involve fixed multiplication by ten, for example, it *may* be more efficient to shift left twice, add in the original, and shift left again. On the other hand, if the math library is already included elsewhere, there may be no code space savings and only minimal speed improvement.

ROTATE AND SHIFT DON'T MODIFY VARIABLE: Don't think a rotate or shift operator changes the variable shifted. Nothing changes unless the result is assigned back to the thing shifted, as in X=SHL(X,3); or x<<=3;.

ple bytes are needing to be rotated, the outgoing bit moves into the carry to be ready to rotate into the *next* byte. PL/M supports the rotate with R O R and R O L.

Both C and PL/M support the **shift**. A shift is a rotate that always fills 0's on the incoming end and discards any bits falling off the other end. A shift of 8 bits on a BYTE/char variable will always give 0.

ASSIGNMENT OPERATORS

Unique to C is a shorthand representation for modifying a variable and reassigning the result back to the original variable. Normally one would write something like:

$$PORTA = PORTA \& 0xf7;$$

to set the fourth bit up (bit 3 when you start counting with zero) low. Things are shortened to:

$$PORTA \& = 0xf7;$$

using the assignment operator. The assignment operators are valid for $+$, $-$, $*$, $/$, $\%$, $<<$, $>>$, $\&$, $\wedge$, and l.

IDENTIFYING BIT CHANGES

A useful function for logical operators is the identification of *changes*. Suppose you decide to scan a matrix of keys directly rather than use an encoder chip. At regular intervals one matrix column at a time can be driven low. As you drive each successive column, you read in the rows. Keys in the driven column that are pushed come in as zero, and the other rows float high (or are pulled high by the hardware). For a standard 4×4 keypad, information is collected 4 bits at a time. This column scanning could be repeated every 40mSec to give prompt recognition of user inputs and still avoid switch bounce problems. The rest of the program needs only to know of any *changes* to the inputs—new buttons pushed or old buttons released. Logical operators allow changes to be sorted out easily. Figure 2–6 may help to visualize how the logic works here. The logical operations easily and efficiently mark the changed keys. The first part of the example shown in Fig. 2–7 is the scanning of a 4×4 matrix. The high **nibble** (4 bits) is the column outputs, and the low nibble is the rows read back. The result is assembled into two bytes. The last part of the program involves

NEGATIVE LOGIC: It is easy to be confused by negative logic inputs. It is especially important to test any logic processing with a simple example—either by hand or with a program simulator.

previous (old) reading	0	0	1	0	1	1	0	1
most recent (new) reading	0	1	0	0	1	1	0	0
old exclusive-or new	0	1	1	0	0	0	0	1
new pushes (old ^ new & new)	0	1	0	0	0	0	0	0
releases (old ^ new & old)	0	0	1	0	0	0	0	1

Figure 2–6   Logic to Detect Key Changes

comparing the new reading with the previous reading to test for new pushes and new releases. The software was written for hardware that inverts, which makes the logic more intuitively pleasing by driving and reading back ones. Notice how short the C program is. Also, notice that as it is written it is not as intuitively understandable as the PL/M. In Chapter 5, this program is revised as a routine shown in Fig. 5–6.

ARITHMETIC OPERATORS

Add, subtract, multiply, and divide are the basic math operations supported directly in the hardware of most microcontrollers, although the latter two were not in the earliest machines. Multiply and divide can also be done through successive addition/subtraction or through shift-and-add/subtract routines.

It is perhaps deceptive to say that all four are *supported* on the 8051 family. The processor supports them only for (unsigned) bytes. Some of the 16- and 32-

ACCURACY: Many operations in embedded control have very specific limits to the range and precision of numbers. Look very closely at what accuracy you *really* need. If the input is 8 bits, perhaps 8-bit math will suffice if you can be sure intermediate results won't overflow.

MATH VS. LOOKUP: Especially with nonlinear transformations (say the speed of a pump vs. flow rate), it may be sufficient to use a lookup table (and linear interpolation if necessary) rather than the complex math arising out of some curve-fitting algorithm. Anything that can be precomputed and put in a table will run much faster than a multibyte math operation.

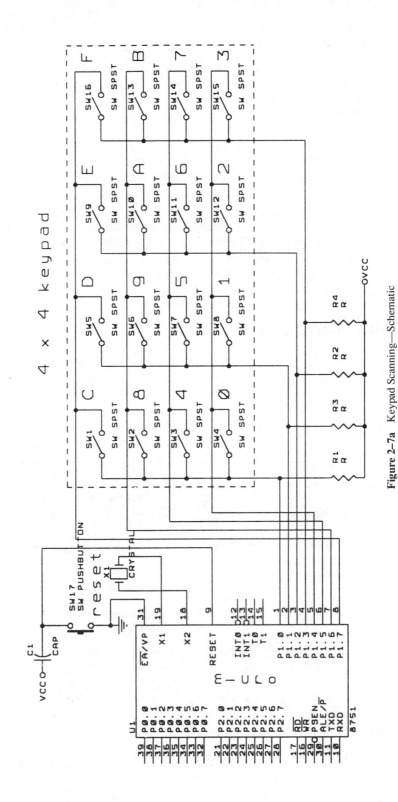

Figure 2–7a Keypad Scanning—Schematic

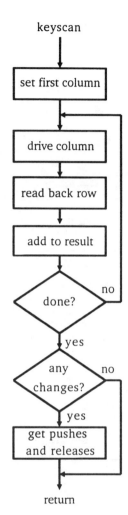

Figure 2–7b   Keyscan Flow Chart

bit processors directly support unsigned integer (word) math, but even there it has been common to rely on a math co-processor. Because most of the 8051 family directly supports only byte operations, *groups* of assembly language instructions handle larger values (there are at least two in the 8051 family that have an added co-processor on-chip for large variables, but in those cases the co-processor is essentially a separate hardware device rather than an extension to the instruction set). For **unsigned int (WORD)** variables, C and PL/M compilers will either code the necessary instructions in-line or add in the necessary procedure/function calls to a library routine. Complete C compilers must have library functions for double precision, signed, and floating-point variables, but PL/M 51 users have to revert to assembly language if handling such variables is demanded by the application.

```
1   #include <absacc.h>
1   #define PORTA XBYTE[0xffc0]
2
    unsigned old,new,push,rel,temp;
    unsigned char clmn_pat;

    void main(void){
3   for (clmn_pat=0x10; clmn_pat<>0;clmn_pat<<1){
    PORTA=PORTA & clmn_pat;
    new=(new<<4) | (PORTA & 0xf);
    }
4   if (temp=new ^ old)>0){
    push=temp & new;
    rel=temp & old;
    }
    }
```

[1]This is an alternate way to define the external ports where you do not want to link in a separate assembly module. The header file includes the definition of the XBYTE function.

[2]This program is somewhat less readable than the PL/M to someone unfamiliar with the use of the special symbols.

[3]The for loop is discussed in the next chapter. By looking for the pattern (clmn_pat) becoming zero, this will loop four times. After the fourth shift left, the 1 bit will "fall off," and the pattern will have all 0's. A **shift/>>** fills 0's instead of bringing bits around as an assembly language rotate would do.

[4]C permits "**embedded assignments**," so the if test includes the filling in of a value for temp. The use of temp could be avoided, but there might then be more code produced because the logical operation would have to be carried out three times. Notice the difference between the = here and the ==, which would only have done the test for equality without assigning any values.

Figure 2–7c   Keyscan Program—C

```
PORTA equ 0ffc0h   1
dseg
    old: ds 2
    new: ds 2
    push: ds 2
    rel: ds 2
cseg
    scan: mov dptr,#PORTA
2   mov r0,#00010000b   ;set column 1
    loop: anl r0,#11110000b ;4 bits low
    mov a,r0
    movx @dptr,a      ;drive column
    movx a,@dptr      ;get row reading
    mov r2,#new       ;store location
3   lcall shift       ;store nibble
    mov a,r0
```

```
      mov a,r0
      rl a   ;next column—left by one
      mov r0,a
      jnz loop      ;not done scanning
      mov a,new     ;test for changes
      xrl a,old
      jnz changed
      mov r0,a
      mov a,new+1
      xrl a,old+1
      jz done       ;no changes—go on
changed: mov r1,a
      mov a,new ;get new pushes
4     anl a,r0
      mov push,a
      mov a,new+1
      anl a,r1
      mov push+1,a
      mov a,old ;get new releases
      anl a,r0
      mov rel,a
      mov a,old+1
      anl a,r1
      mov rel+1,a
done: mov old,new ;update old
      mov old+1,new+1
      ljmp scan ;start over
3
shift: mov r1,#4 ;low 4bits - >16bits
s2: clr c
      rlc @r2+1
      rlc @r2
      djnz r1,s2
      orl acc,#0fh
      orl @r2,acc
      ret
end
```

---

[1] s c a n is a "complete" program but rather pointless since nothing is done with the results of the key scan. Chapter 5 goes into the details of subroutines, which would be a better use for this program. The s h i f t part *is* a subroutine since it ends with r e t and is called from the main program.

[2] The main program, s c a n, involves going four times around l o o p in order to drive the four columns and read back the four rows each time. The result is put into the two-byte storage called n e w.

[3] This routine is used to gather the 4-bit readings into a 16-bit (2-byte) result. Notice how the r l c allows the low-byte bits to carry across into the high-byte. When the previous value has moved over by 4, then the lower 4 bits are masked and the newest 4-bits are o r-ed in.

[4] The logical operations described above to sort out new pushes and releases begin here.

Figure 2–7d   Keyscan Program—ASM51

```
  SCAN:DO;
   DECLARE (OLD,NEW,PUSH,REL,TEMP)
1    WORD AUXILIARY;
   DECLARE (I,COLUMNPATTERN)
     BYTE AUXILIARY;
   DO WHILE 1;
2  COLUMNPATTERN=00010000B;
   DO I=0 TO 3;
     PORTA=PORTA AND COLUMNPATTERN;
     NEW=SHL(NEW,4) OR (PORTA AND 0FH);
     COLUMNPATTERN=SHL(COLUMNPATTERN,1);
   END;
3  TEMP=OLD XOR NEW;
   IF TEMP0 THEN DO; /*SOME CHANGE*/
     PUSH=TEMP AND NEW;
     REL=TEMP AND OLD;
   END;
   OLD=NEW;
   END;
  END SCAN;
```

[1]Notice how much more readily understandable a high-level language can be.

[2]Binary notation gives a better picture of the *function* of the constant—here to show that only one column is being driven. By the way, it is often better to drive the desired line *low* and read back *lows* for pushed buttons because TTL is more effective for *sinking* current with pull-up resistors on the lines. See Appendix B for details.

[3]The use of TEMP is to avoid repeating the logic operation three times. If the values were only bytewide and were kept in off-chip RAM, it might well be better to recompute TEMP each time.

Figure 2–7e   Keyscan Program—PL/M

An example in Chapter 6 (Fig. 6–8) shows a set of assembly language double-precision math routines linked with C and PL/M if the idea is attractive to you. Most C programmers are quite content to rely on the libraries supplied with the compiler unless severe speed constraints exist.

C and PL/M automatically do **type conversion** (expanding bytes to

TRUNCATING DUE TO VARIABLE SIZE: By adding 2 one-byte variables and assigning the result to a 2-byte variable without doubling/casting one of the 1-byte variables, the compiler may truncate the result to a byte at the addition and *then* promote it to 2 bytes when doing the assignment.

```
  unsigned int a,b;
  unsigned char c;
1 a=b+c;

  unsigned int a;
  unsigned char b,c;
2 a=b+(unsigned int)c;
```

[1]The program takes b, treats c as though there were 8 high-order zeros attached, does the math, and assigns the result to the 2 bytes of a. If a were an unsigned c h a r, then the 2-byte result of the math might have the upper byte discarded (even if that included nonzero bits).

[2]Where b and c are bytes and a is a i n t, the math could be carried out with single byte precision, and the carry bit might be lost *before* the result is converted to a 2-byte quantity. Some C compilers for other processor families **promote** (change—enlarge) all c h a r variables to i n t before doing any math anyway. This **type cast** could force other conversions as well: floating, double, etc.

Figure 2–8a   Type Conversion—C

```
  DECLARE (A,B) WORD;
  DECLARE C BYTE;
1 A=B+C;

  DECLARE A WORD;
  DECLARE (B,C) BYTE;
2 A=B+(DOUBLE(C));
```

[1]The program takes B, treats C as though there were 8 high-order zeros attached, does the math, and assigns the result to the 2 bytes of A. If A were a byte, then the 2-byte result of the math would have the upper byte discarded (even if that included nonzero bits).

[2]Where B and C are bytes and A is a W O R D, the math could be carried out with byte precision, and the carry bit might be lost before the result is converted to a 2-byte quantity. The D O U B L E operator forces C to be a 2-byte quantity so that 2-byte math will be used. That saves the carry bit between the 2-byte operations.

Figure 2–8b   Type Conversion—PL/M

word/integer, etc.). For example, if one adds a byte to a word, the result will be a word as shown in the first part of Fig. 2–8. Both C and PL/M have an operation to *force* a variable to a different type, as shown in the second parts of Fig. 2–8.

The 8051 family of processors has a very limited set of built-in arithmetic instructions. As with the logical operators, they work only on internal memory and must involve the accumulator to hold the result. None of them work with more than single bytes at a time (an *8-bit* processor), although multiply and divide

DEDICATED MATH ROUTINES: Sometimes running speed is at a premium, and the application needs only very specific math operations. Then it can be more efficient to write dedicated assembly routines to replace the more general library functions. For example, a multiply might be needed, but it will always be a multiply by 5. Or, an 8 X 16 multiply might be needed but you can guarantee in your application that the result will never exceed 16 bits. Since the library function may carry out a 16 X 16 multiply and truncate the result, a much shorter version which is less general may be more efficient. Chapter 6 on modular programming discusses more details of including these custom routines. There are parameter passing conventions that must be observed in any assembly module that works with high-level languages.

CONFUSING REMAINDER WITH FRACTION: Don't think the 8- bit divide with a 16-bit result leaves the whole number and the fraction. Actually the divide leaves the *remainder* rather than the fraction.

do leave 16 bits of result. There are machine instructions that involve the carry bit, making it straightforward to process any number of bytes one at a time, with the carry holding over between bytes. Because C and PL/M handle 2-byte and larger variables in software (usually from libraries) instead of by user programming, by looking at assembly language examples you can imagine what the general approach in the high-level libraries may be.

Figure 2–9 shows the addition or subtraction of two unsigned 16-bit numbers. Obviously this is trivial in C or PL/M.

For a second example, 8 × 8 multiplication and division are shown in Fig.

```
unsigned int x,y;
x = x+y;

unsigned int x,y;
x = x-y;
```

Figure 2–9a   2-byte Addition and Subtraction—C

```
a16:    mov  a,r3
1       add  a,r5
        mov  r3,a
        mov  a,r2
2       addc a,r4
        mov  a,r2


s16:    mov  a,r3
3       clr  c
1       subb a,r5
        mov  r3,a
        mov  a,r2
2       subb a,r4
        mov  a,r2
```

[1]A problem with assembly language is the number of decisions about where to keep things. There are pointer-based instructions (based on r0 or r1) and direct (to internal locations) instructions. Here, though, let's assume the first number is in r2/r3 (high byte/low byte) and the second number (to be subtracted *from* the first in the second case) is in r4/r5. Suppose the result is to be overlaid on top of r2/r3. First add (or subtract) the two low bytes, being sure the carry bit does not come in at the bottom. In a more-than-two-byte routine, it might be better to clear the carry first and loop with the addc instruction.

[2]Be sure to include the carry bit without doing anything to destroy it in between. Add the second two bytes. If there is a carry when done, the result is more than 16 bits. What to do depends on what should happen next. If you need a fast, efficient routine, make sure that an overflow *can't* happen by checking all the possible values that could come to the routine in the final application. In more general routines, set an "error" bit variable, and have the calling routine check the flag before using the results.

[3]There is no instruction to subtract without including the carry bit, so it *must* be cleared first.

Figure 2–9b   2-byte Addition and Subtraction—ASM51

```
DECLARE (X,Y) WORD;
    X = X+Y;


    DECLARE (X,Y) WORD;
X = X-Y;
```

Figure 2–9c   2-byte Addition and Subtraction—PL/M

2–10. The only high-level language issues relate to ensuring a 16-bit result is produced.

As an example of the mechanics of handling bigger numbers, the routine of Fig. 2–11 does 4-byte add. The mechanics are the same as earlier, but a counter tells when the fourth byte is done. The values are on an artificial stack made up of STKa and STKb, with the msb of each at the lowest address and the lsb at

FORGETTING OVERFLOW: By omitting the cast or double when a result needs to be larger than either of the variables involved, it is quite likely the compiler will not promote the result until after the math is done and the upper part of the result will be lost.

STK+4. It is a part of a family of routines discussed in the section of Chapter 6 that discusses mixed languages. There the routines to push this stack, take in WORD/integers and extend them to 4 bytes, multiply, and divide are also given.

At this point it would be possible to go into multibyte math, including signed and floating variables, but the availability of similar functions in C libraries as well as the infrequent need of such variables in efficient controller programming makes such math of little interest to most readers and will be omitted.

LOGICAL OPERATORS

One of the most easily confused features of C is the distinction between bitwise operators and logical operators. As mentioned already, bitwise operators modify the values of individual bits of a variable. Logical operators instead produce a "true" or "false" answer about the relationship between variables or expressions. They are used extensively in the conditional tests of the looping and branching of

```
    unsigned char x,y;
1   unsigned int z;
    z = (unsigned int)x * y;

    unsigned char x,y,a,b;
2   a = x / y;
3   b =  x % y;
```

[1]The cast is used to ensure the production of a 16-bit result for the multiply. It may not be strictly necessary with the compilers in either language, but emphasizes the fact that more than 8 bits are possible and that the result is generally put into the same type as the two initial variables. What would happen with mixtures of fixed and floating-point numbers is not certain, but it is not likely that the compiler would *automatically* promote them to fixed.

[2]The integer portion of the division is put into a. Remember that it is not a number with a fractional part.

[3]The % operator obtains the *remainder;* not the fractional part. The treatment would be quite different for floating-point variables.

Figure 2–10a   1-byte Multiply and Divide—C

```
var segment data
rseg var
1   term1:ds 1
    term2:ds 1

mult segment code
rseg mult
    mov r0,#term1
2   mov r1,#term2
    mov a,@r0
    mov b,@r1
3   mul ab
    mov r7,a
4   mov r6,b
end


var segment data
rseg var
1   term1:ds 1
    term2:ds 1

div segment code
rseg div
    mov r0,#term1
2   mov r1,#term2
    mov a,@r0
    mov b,@r1
5   div ab
    mov r7,a
6   mov r6,#0
end
```

[1]Here the instructions exist directly, but the in and out moves are of some interest. Assume r0 points to the first (or numerator) term, and r1 points to the second (or divisor) term. For variety, suppose the result (16 bits) is to be left in r6/r7 in the usual high/low order. The variables are defined as relocatable (assigned at link/locate time) variables called term1 and term2. If you want specific locations, define them with the equ directive instead. Chapter 6 discusses these details more.

[2]Notice that r0/r1 holds the address (not the value) of the variables. That way the pointer-based moves that follow will get the right values.

[3]The multiply produces a 16-bit result with the high part in the b register. It also sets a flag (the carry) if the result exceeds 1 byte.

[4]Technically the instruction moves to r6 from any direct internal *address,* and b is not an address. It may be the case that the direct address for the b register, 0f0h, will need to replace b if your assembler does not do it automatically.

[5]The quotient is left in a (the integer part).

[6]The integer remainder (*not at all the same* as the fractional result) is left in b, so if further division is to be done, it can follow. For example, 20 divided by 7 would leave 2 in a and 6 in b. One could then rotate b left by one (multiply by 2) to be 12, carry out the division by 7 again to fill the 1/2 bit place, shift again to fill the 1/4 bit place, etc. This approach is no more cumbersome than the shift-and-subtract type of division, but it would run slightly slower due to the time required for the div instruction.

Figure 2-10b   1-byte Multiply and Divide—ASM51

```
   DECLARE (X,Y) BYTE;
   DECLARE Z WORD;
1  Z = DOUBLE(X) * Y;

   DECLARE (X,Y,A,B) BYTE;
2  A = X / Y;
3  B = X MOD Y;
```

[1]The DOUBLE is used to ensure the production of a 16-bit result for the multiply. It may not be strictly necessary, but emphasizes the fact that more than 8 bits are possible and that the result is generally put into the same type as the two initial variables.

[2]The integer portion of the division is put into a. Remember that it is not a number with a fractional part.

[3]The MOD operator obtains the *remainder*, not the fractional part.

Figure 2–10c   1-byte Multiply and Divide—PL/M

```
   unsigned long stka,stkb;
1  stka += stkb;
```

[1]Notice how simple the C version looks compared to the other language examples. That isn't to say it wouldn't produce a large amount of code, but the details are left to the compiler. Actually the example used 30 bytes.

Figure 2-11a   4-byte Addition—C

the next chapter. The problem is that bitwise and logical "or" and "and" operators have the *same* name and there is the assignment "equal" and the logical test for "equal." In C the logical equal is == (a double equal) and the logical "or" and "and" are && and ||. Particularly problematic is the fact that the compiler can correctly accept either (with radically different results) in many expressions!

PL/M simply uses the same words. Because embedded assignments are not allowed, the compiler has no trouble, for example, sorting out the assignment "=" from the logical test "=".

"True" and "false" in a logical sense differ for C and PL/M. In C, if a value of a variable in a logical test is greater than zero, it is "true." In PL/M a variable is true or false based on whether or not there is a 1 in the least significant bit (lsb). A resulting shortcut for C is the omission of a "greater than zero" in a test for a bit: if (PORTA | 0x40) ...; can suffice for: if ((PORTA | 0x40)>0) ...;.

PRECEDENCE

How do compilers interpret lines with several operations? For example, A + B * C *could* mean, going left to right, that A is added to B and *then* C is multiplied by

```
adding segment code
store segment data
rseg store
1   STKa: ds 4
    STKb: ds 4
    STKc: ds 4
rseg adding
dadd:mov r0,#STKa+3 ;lsb of A
    mov r1,#STKb+3 ;lsb of B
2   mov r2,#4 ;4 bytes to process
    clr c ;no carry into first addition
dad1:mov a,@r0
    addc a,@r1 ;A+B
    mov @r0,a ;save in A
3   dec r0 ;move to next byte
    dec r1
4   djnz r2,dad1 ;4 times
    mov r0,#STKc
    mov r1,#STKb
5   acall qmov ;move C over B
    ret

qmov:mov r2,#4
qmo1:mov a,@r0
6   mov @r1,a
    inc r0
    inc r1
    djnz r2,qmo1
    ret
end
```

[1]The values will be drawn from S T K a and S T K b much like the previous example's use of t e r m 1 and t e r m 2. The only registers available for pointers are r 0 and r 1.

[2]It is more code efficient to use a counter rather than repeat the loop contents four times.

[3]The program decrements the pointers because the math must go from least to most significant, and the stack starts with msb at the lowest address.

[4]This powerful instruction decrements the counter and conditionally jumps in one instruction. When the addition of the fourth bytes is done, it goes on.

[5]Since A + B has been (arbitrarily) defined to destroy A and B, leaving the result in A, the last part moves the furthest in (top? bottom?) term back one. If A or B is to be saved as in a stack-oriented math chip, it would first have to be duplicated on the stack. This is not the place to discuss stack-oriented math and reverse Polish calculations, but this approach is consistent with that approach.

[6]This subroutine is an example of breaking jobs up into pieces, as will be discussed in Chapters 6–8. Be careful when moving blocks or strings that the move doesn't overwrite the old values before they are all moved. For example, if the stack had 4 terms, then C should go down to B *before* D goes down to C.

Figure 2–11b   4-byte Addition—ASM51

```
DECLARE STKA(4)BYTE,STKB(4)BYTE,I BYTE;
DECLARE TEMP WORD;

DO I=0 TO 3;
1   TEMP=DOUBLE(STKA(3-I))+STKB(3-I);
2   STKA(3-I)=LOW(TEMP);
    IF HIGH(TEMP)>0 AND (I<3) THEN
    STKA(2-I)=STKA(2-I)+HIGH(TEMP);
    END;
```

[1]The complex index calculation is used because the iterative d o cannot index down and it was desired to have the 4-byte arrays arranged high-to-low. The addition must begin with the least significant byte in order to have any carries move up.

[2]This sequence is necessary to take the carry to the next higher byte. It *might* be possible to compile in such a way that the carry would remain and could be used next, but it would be necessary to compile with the *code* option to be sure of this. Clearly, PL/M is not designed for this sort of math.

Figure 2–11c   4-byte Addition—PL/M

IGNORING PRECEDENCE IN PORT MASKING: See the upper part of Fig. 2–12. Since the bitwise operator AND/& has a *lower* precedence than the relational operator >, the first operation is really for testing whether 20H/0x20 is greater than zero. The result is *always* true (which evaluates to a 0xff or a 1). Then the port is either not masked at all or masked for the lsb (bit 0) rather than bit 5. The final project using this error will "mysteriously" freeze or race on, depending on what happens with other bits or may totally ignore bit 5. The correct alternative shown in the lower part of Fig. 2–12, where the parentheses force the masking *before* the test for a non-zero result.

```
while (porta & 0x20  > 0) {};

while ((porta & 0x20)  > 0) {};

DO WHILE PORTA AND 00100000B  > 0;
END;

DO WHILE (PORTA AND 00100000B)  > 0;
END;
```

Figure 2–12   Common Precedence Error

the result. Actually, it means that B is multiplied by C *before* A is added; as is normal in algebra, multiplication is given higher precedence than addition. Using parenthesis, A + (B * C), avoids the problem and eliminates any doubt because the expressions nested within a () pair must be evaluated *before* the result can be used outside. In assembly there is only one operation per line and a top-to-bottom flow, so there are no precedence issues. In both C and PL/M, the precedence rules are those of algebra, but mixed math and logic expressions can be misleading.

Figure 2–13 presents the precedence rules in tabular form. Some of the operators, particularly in C, will not be discussed until later chapters. There are several minor precedence differences between the languages. Notice that the NOT is much lower in precedence than the ~ operator. Also, C makes a precedence distinction between ! =, ==, and the rest of the relational operators. It also makes a distinction between logical operators (&&) and bitwise operators (&) that PL/M does not do. If you write *understandable* code, these minor points of precedence will never come up because liberal use of parentheses will avoid any confusion!

PL/M Operator	C Operator	Order of Evaluation
( )	( ) [ ] –>.	l to r
– . +	! ~ ++ — (type)	
	* & sizeof	r to l
* / MOD	* / %	l to r
+ – PLUS		
MINUS	+ –	l to r
	<< >>	l to r
< <= <>		
= >= >	< <= > >=	l to r
	==!=	l to r
NOT		l to r
AND	&	l to r
OR XOR	^	l to r
	|	l to r
	&&	l to r
	||	l to r
	?:	r to l
	= += -= etc	r to l
		l to r

Figure 2–13   Precedence of Operators

REVIEW AND BEYOND

1. Why are BASIC examples not included in this book?
2. Which variable types does the 8051 family directly support? Which ones require additional software routines?

3. What are the three main memory spaces in the 8051 family? How are the two types of RAM space different?
4. Are the line-feed and return characters significant in C and PL/M programs? Explain. Discuss indentation and features that might improve readability in a program listing.
5. What is a *type cast* in C? What is an assignment operator?
6. Write a piece of program to compare a new 8-bit port reading with a previous value. Then produce a number having 1's only where the reading has 0's that were not present in the previous reading.
7. Explain some of the costs of using floating-point math in C.
8. Which operators have the highest precedence in C and PL/M?

Chapter 3

Looping and Branching

DECISIONS

If computers did only the basic operations described in Chapter 2, there would be very little usefulness to them. Their power lies in the ability to make decisions:

Based on the condition of a switch, turn on the water or not.

If this operation has been repeated 22 times, then go on and do the next operation.

Keep on checking for the signal telling you that the speech chip can take in the code for the next word.

All these are examples of **decisions** that a microcontroller is routinely required to make. Based on the decision test, the program flow will **loop** (go back on itself) or **branch** (go in one of several possible directions).

FLOW CHARTS

Before proceeding, a word is in order regarding **flow charts**. Basic parts of flow charting will be illustrated as this chapter develops. Those used throughout *this* book may be different from flow charts you have been exposed to before. The flow charts here are not intended as a *detailed* picture of the program, but rather a quick *overview* of the method of solution. Just as Chapter 5 will argue that program code should be made up of pieces no longer than a page, so flow charts should fit on a single page. There are horror stories relating to many-page flow charts with various paths to points several pages away. Although perhaps technically correct, such flow charts don't help anyone understand what is going on in the program. To encourage *understandable* flow charts, when they become too complex to fit on one page, simplify them (and the program the flow chart represents) by assigning pieces to subroutines (discussed further in Chapter 5). The *details* of those subroutines can be quite properly expanded in *separate* flow charts on other pages. Thus, the *main* flow chart will always give an overview of the pieces of the program. If you need more detail as to the method of solution, go to the appropriate subroutine.

All flow charts should talk about things by their functional names and should generally *not* refer to specific variable names. For example, consider a routine to wait for a button to be pushed. As shown in Fig. 3–1, it is better to have a block like "WAIT FOR START BUTTON," rather than a block like "LOOP TILL P1.3=0." Better yet might be a loop shown as the third choice.

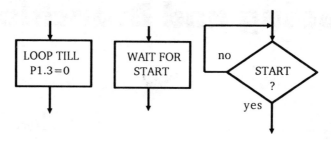

Figure 3–1   Functional Flow Charts

The program pieces, which follow, all include examples of suggested flow chart symbols. Computer science, particularly in the area of data processing, has a much more elaborate set of symbols, but most of them relate to complex subsystems as parts of the program flow and are not considered appropriate here. There are several approaches to iterative loops in flow charts, which may be more efficient, but they will not be covered here.

STRUCTURED LANGUAGE

Both C and PL/M are **structured languages** in the sense that there is a rigid set of language constructs that do not allow criss-crossed program flow. Viewed from another perspective, a structured language is one that never jumps into or out of a routine without saving or restoring the stack and any other pertinent registers. With structured programming, aside from interrupts which are a special case discussed in Chapters 8 and 11, the stack cannot be corrupted with any acceptable commands. Structured programming rules *can* be followed with assembly and BASIC. QUICKBASIC is an inherently structured language.

The basic element of structured languages is the **block**. It is a section of program where the flow enters at *only* one place and leaves at *only* one place. There can be no "sneaking" into the middle sometimes or leaving partway through the block. Figure 3–2 shows an example of a simple block in C and PL/M. The first box shows the representation used in flow charts. The other two boxes are the PL/M and C form of the block. Assembly does not define a block

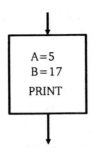

Figure 3–2a Blocks

```
{
  a=5;
  b=17;
  print;
}
```

Figure 3–2b Basic Block Structure—C

```
DO;
  A=5;
  B=17;
  CALL PRINT;
END;
```

Figure 3–2c Basic Block Structure—PL/M

per se, although a straight line piece of program with no jumps or labels can be considered a block in that it has only one entry and one exit point.

Remember that the individual C or PL/M statements do not have to be on separate lines because the line feed is not significant to the compilers. Programs can be made "shorter" and "wider" that way. This is particularly useful where several short expressions have a closely related function. The block of Figure 3–2 could be written as shown in Fig. 3–3.

```
{
  A=5; B=17; print();
}

{A=5; B=17; print();}
```

Figure 3–3a Alternate Styles in a Block—C

```
DO;
  A=5; B=17; CALL PRINT:
END;
```

Figure 3–3b   Alternate Styles in a Block—PL/M

BRANCHING CONSTRUCTS

C and PL/M both provide the **goto** statement somewhat corresponding to an assembly language jump. Of course, machine code jumps are the *result* of structured programming, but the structure keeps the stack depth correct and restores variables. A raw jump into the middle of a routine that is subsequently left by a return will pop the stack too much and will return to some unplanned address. In C, the only allowed goto is within the same function to a labeled line. PL/M permits a GOTO to a place at the same depth of nesting (the same DO block) or to an outer level of an enclosing block. It is not permissible to do a GOTO out of a procedure into another procedure. Probably a goto is never necessary. The only commonly accepted uses are to get out of a routine when an error condition comes up—particularly when the errors may come at different levels of nesting and at different points in loops—or to cut short a search when the goal is reached. Both can be done without the goto, but it is sometimes more understandable with the goto included.

C has an additional set of seldom discussed instructions for terminating loops—the **break;** and **continue;** directives. The former is frequently used with the switch construct discussed later, whereas the latter can be used to cause flow to drop out of a while loop.

The sections that follow discuss branching and looping constructs. The assembly language examples show a series of instructions to accomplish the same *function* as the high-level languages. You will notice a strong similarity between C and PL/M, although C has several more choices and more flexibility.

IF/THEN/ELSE

The most basic decision or branch in program flow is the **if/then/else** block. In assembly this is done with the jz, jnz, jb, jnb, jc, and jnc instructions. All of the conditional branching for the 8051 consists of jumps—there are no conditional calls directly available. There are also several *unusual* looping instructions discussed in the next section. The jz / jnz instruction works on the basis of the contents of the accumulator at that moment—there is no zero flag as with the 8080 family. Caution is necessary with the carry bit to be sure an instruction between the carry setting place and the actual jump test doesn't affect the carry. An example is given in Fig. 3–4.

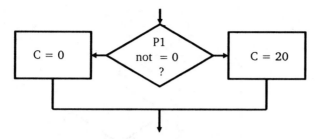

Figure 3–4a Decision Blocks—Flow Chart

```
if(P1!=0) c=20;
else c=0;
```

Figure 3–4b Decision Blocks—C

```
     mov  r0,#c
          mov  a,P1
          jz   x
y:        mov  @r0,#20h
          sjmp z
x:        mov  @r0,#0
z:              . . . .
```

Figure 3–4c Decision Blocks—ASM51

```
IF (P1<>0)THEN C=20;
ELSE C=0;
```

Figure 3–4d Decision Blocks—PL/M

The only difference between the two high-level languages is that C allows assignments *within* the test expression, called **embedded assignments**, whereas PL/M 51 (unlike PL/M 80) does not. This is where the ++i as opposed to the i++ can be significant—does the variable get incremented *before* or *after* doing the test? Obviously, that sort of distinction can easily be missed by the casual reader. The e l s e is optional, and the expressions can be blocks in themselves, as shown in Fig. 3–5.

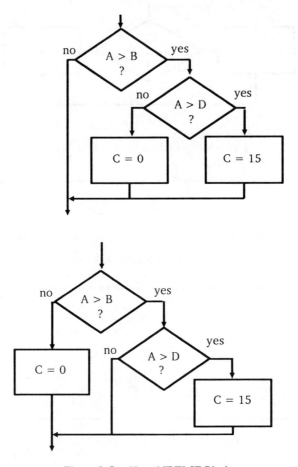

Figure 3–5a   Nested IF/ELSE Blocks

CONDITIONAL OPERATOR

An expression unique to C is the conditional operator. It is a shorthand for an if/else decision where the two choices simply assign a different value to a variable. It is a test where the true condition assigns the first value and the false condition assigns the second value. Figure 3–5d shows the first case of Fig. 3–5b using this operator.

SWITCH/DO CASE

Besides the simple branching construct of the i f / e l s e, there is the somewhat less used **switch/DO CASE** construct. It allows for branching out many ways

```
   if (a>b) {
1  if (a>d)  c=15;
   else c=0;
   }

   if (a>b)
    if (a>d)  c=15;
   else c=0;

   if (a>b)   {
    if (a>d)  c=15;
   }
   else c=0;
```

¹Note that the nesting of the i f's ties the e l s e with the *most recent* i f unless the block structure defines it otherwise. In Fig. 3–5, the first and second blocks are identical, though the (misleading) indentation suggests that the second example will set C=0 if A is not greater than B. If that is what you want, then in the third example the necessary { } block shows how to make the e l s e apply to the first i f.

Figure 3–5b   Nested IF/ELSE Blocks—C

```
   IF (A>B) THEN DO;
1  IF A>D THEN C=15;
   ELSE C=0;
   END;

   IF (A>B) THEN
    IF A>D THEN C=15;
   ELSE C=0;

   IF A>B THEN DO;
    IF A>D THEN C=15;
   END;
   ELSE C=0;
```

¹Note that the nesting of the I F T H E N's ties the ELSE with the *most recent* I F T H E N unless the block structure defines it otherwise. In Fig. 3–5, the first and second blocks are identical, though the (misleading) indentation suggests that the second example will set C=0 if A is not greater than B. If that is what you want, then in the third example the necessary D O / E N D block shows how to make the E L S E apply to the first I F T H E N.

Figure 3–5c   Nested IF/ELSE Blocks—PL/M

```
if (a>b)  c = (a>d)?15:0;
```

Figure 3–5d   Conditional Operator—C

FOOLING YOURSELF WITH INDENTATION: It is possible to assume the last t h e n or e l s e applies to the group of lines that come after when the operation applies to only the first line. Unless the group is surrounded by D O / E N D or { }, the first element of the group will be alone and the rest of the group will be part of the resumption of the general flow.

```
if (a>b)                                    IF (A>B) THEN
  c=25;                                        C=25;
  g=a+b;                                       G=A+B;
  P1=7;                                        P1=7;
```

If you program carefully, this will never be a problem. If not, many forms of this error will cause compiler errors, but carelessly fixing the errors by inserting D O / E N Ds or { }s until the errors "go away" may give the wrong result.

based on the value of an expression. A string of i f / e l s e constructs can do the same job, but sometimes a s w i t c h / D O C A S E makes a program more understandable. In its simplest form, the branching is based on an integer as shown in Fig. 3–6.

LOOPING CONSTRUCTS

The branching constructs all carry the flow on *forward* (except for some uses of GOTO). With the *looping* constructs it is possible to *repeat* things. In assembly,

INDEX OUT OF RANGE IN PL/M: In the example above, a value of 4 for K would cause a jump to an undefined location. (See note 3 for Fig. 3–6c.)

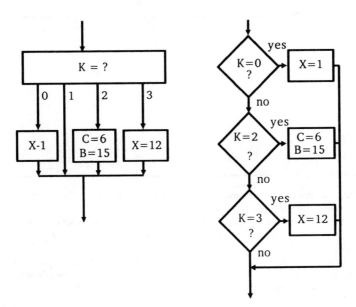

Figure 3–6a Basic Case Branch

```
switch (k){
1  case 0:
2    x=1; break;
   case 2:
     c=6;
     b=15;
     break;
   case 3:
     x=12;
3  default:
     break;
 }
```

[1]Each case is tested against the *constant* value following the case, but it is not necessary to have them in any particular order. Any missing cases will cause no action.

[2]Program flow will *fall through* to the next case unless there is a **break**. In other words, without the break, case 0 would execute cases 2 and 3 as well. It is unnecessary to put braces around the cases, since by default, execution goes until the b r e a k is encountered.

[3]The d e f a u l t that says that if *none* of the listed cases match, then do the default. It is optional. In any case, all flow resumes after the }, and values of k except the ones specified in the other cases will safely fall out as well. The default in this example is totally unnecessary since a case not in the list will automatically exit in any case—it would make more sense if there were some alternate processing if the valid cases were missed.

Figure 3–6b Basic Case Branch—C

```
   DO CASE K;
1  X=1;
   ;
2  DO;
     C=6;
     B=15;
   END;
   X=12;
3  END;
```

---

[1]The cases must be in numeric sequence beginning with zero. Any integer missing in the progression must have a null (;) expression, as is shown for case 1.

[2]Only the specific case is executed—flow does not fall through to the next case. A DO block can make a case include several expressions (or a whole long program piece with calls to other procedures).

[3]The END terminates the DO CASE, but there is no checking for cases beyond 3 in this example. If K should happen to be 5, there would be a disaster—a branch to an undefined location. It is wise to put a test on K ahead of the DO CASE such as IF K<4 THEN DO.

Figure 3–6c   Basic Case Branch—PL/M

FORGETTING THE BREAK IN C: Without the break, the program will flow down through the following cases. Those familiar with PL/M might assume each case would terminate when done and resume after the switch block, but the streams of flow for successive cases will add together into a river if you omit the break. This can be powerful for some applications, but a nuisance in other cases.

of course, the jump instructions have the same effect. The high-level looping constructs are discussed next.

WHILE LOOP

One looping construct, shown in Fig. 3–7, is the **while** block. The program flow continues in the loop until some condition test fails. One form uses the test *first*, entering the block *only* if the test passes. If not, then the flow skips over the block and continues with the first statement *after* the block. A second form, available only in C, does the test at the *end* of the block to decide whether to go back and

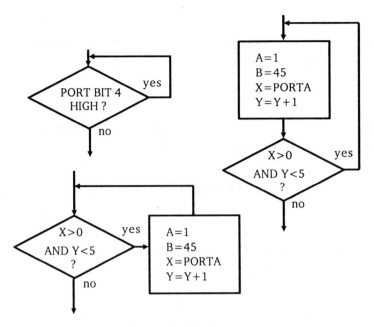

Figure 3–7a While Loops

```
1   while ((P1 & 0x10)==0){}

    while (x>0 && y++==5){
2       a=1;
        b=45;
        x=P1;
    }

    do{
        a=1;
3       b=45;
        x=P1;
    }
    while (x>0 && y++==5);
```

---

[1]By testing *until* the fifth bit (bit 4) goes *high*, this loop waits for some signal from a user or external hardware. It is safer to test with a greater than zero rather than an equal to 10H because the latter leaves the possibility that the masking number and the equality test number might get messed up and not agree.

[2]The incrementing of the value of y is simply done by the y++. Remember that the == is different from the =, which would change the value of y strangely.

[3]This block (only directly possible in C) will execute the block once before doing any testing of the values of x and y.

Figure 3–7b While Loops—C

```
     mov dptr,#porta
x:   movx a,@dptr
     anl a,00010000b
1    cjne a,00010000b,x
```

¹By testing until the fifth bit (bit 4) goes high, this loop waits for some signal from a user or external hardware. The cjne is very useful here.

Figure 3–7c   While Loops—ASM51

```
     DO WHILE (P1 AND 10H)=0;
1    END;                    2

     DO WHILE (X>0 AND Y=5);
          A=1;
          B=45;
          X=P1;
          Y=Y+1;
3    END;
```

¹By testing until the fifth bit (bit 4) goes high, this loop waits for some signal from a user or external hardware. It is safer to test with a greater than zero rather than an equal to 10H because the latter leaves the possibility that the masking number and the equality test number might get messed up and not agree.

²The parentheses are crucial due to the precedence. PL/M does not require a parenthesis around the entire expression as is required in C.

³The incrementing of Y must be done inside the loop in PL/M.

Figure 3–7d   While Loops—PL/M

do it again or to go on. Thus, the block is always executed at least once. Unfortunately, the PL/M name for the first is the same as the C name for the second! (Incidentally, in PASCAL the two forms correspond to DO WHILE and DO UNTIL, which may be clearer.) One of the unusual and powerful assembly instructions is for carrying out looping. The cjne instruction compares the first two operands and branches only if they are not equal. It can easily make a loop to test a port until a particular value comes in. Figure 3–7 gives an assembly example of that as well as several more complex examples of the two kinds of loops with C and PL/M.

ITERATIVE LOOP

A second very common looping block structure, shown in Fig. 3–8, is the **iterative** block. It involves either an index counter variable or, as an additional option with C, *any* expression. There are three parts to the block besides the body of instructions to be repeated. First is the initial expression: for PL/M this must be an initial value to assign to the index variable; for C this can also be *any* expression to be done once at the start. Then there is the test for ending the looping: for PL/M it must be the upper value of the index for which looping can continue; for C it can be *any* test whose failure will terminate the loop. Finally, there is the size increment. By default in PL/M it is one and can only be positive. Each time around the loop the increment is applied to the index variable. With C, any indicated operation or expression to be performed after the test before re-entering the block can be here. C can do nearly incomprehensible things with such a construct if you are feeling powerful and secretive!

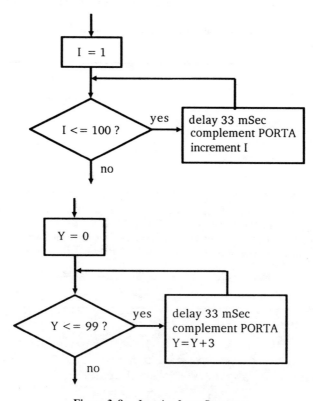

Figure 3–8a   Iterative Loop Structures

```
    for (i=1;i<=100;i++){
    delay(33);
    px=~px;
    }

    for (y=0;y<=99;y=y+3){
    delay(33);
1   px=y;
    }

    for (da=start; status==busy; leds=~leds){
2   delay(33);
    }
```

[1]This example increments y by 3 each time around the loop. The i++ could be put into the second part of the for (;;) expression, and the third part could be left blank. The body of the operation could even be put into the third part! Such embedding can be quite powerful, but also quite confusing.

[2]The last example would use #define to designate the various port and bit assignments. It would start an A-D converter and wait for it to complete the conversion. It would toggle some leds at a rate determined by the delay and appear to the casual reader to do nothing but delay in the loop. For an A-D converter, the next action after the loop ought to be to pick up the completed reading.

Figure 3–8b Iterative Loop Structures—C

```
    mov r0,#100
x:  mov r1,#33
    lcall delay
    mov a,p1
    cpl a
    movx p1,a
1   djnz r0,x
```

[1]One assembly instruction, djnz, is quite powerful for small iterative loops. It can use one of the registers or a specified internal RAM location as the index counter. By its nature, it is designed to decrement by one each time, but if other decrements were desired, additional decrement instructions could be included just before the loop test.

Figure 3–8c Iterative Loop Structures—ASM51

MIDDLE TEST OF FOR LOOP: In C it is easy to confuse the middle term of the loop, which must be *true* to continue in the loop, with a term that is the *ending* point of a counter.

```
    DO I=1 TO 33;
    CALL DELAY(33);
    PX=NOT(PX);
    END;

1   DO Y=0 TO 99 BY 3;
    CALL DELAY(33);
    PX=Y;
    END;
```

[1]This example increments y by 3 each time around the loop.

Figure 3–8d Iterative Loop Structures—PL/M

EXAMPLE: TIME DELAY

One of the common uses for looping constructs is the production of a delay time by nesting loops so that the instruction execution consumes a known amount of time. Examples of such time delay are shown in Fig. 3–9. The time is based on the number of clock cycles, so the delay will depend on the crystal frequency. The 8051 data book lists the number of clock cycles required for each machine instruction. With a 12MHz crystal, 12-oscillator-period instructions take 1 microsecond, and 24-oscillator-period instructions take 2 microseconds. The only exceptions are the mul and div, which each take 4 microseconds. To generalize

```
1   void msec(unsigned int x){
    unsigned char j;
    while (x- -0){
2       for (j=0;j<125;j++){;}
    }
    }
```

[1]This routine can take integer values to produce long delays. It is not precise as written, although it is approximately correct based on an analysis of the assembly showing the most inner loop with j taking 8 microseconds. Different compilers may come out with quite different times and the value of 125 could be adjusted to compensate empirically. As will be mentioned in Chapter 7, delay loops are often an inefficient way to get delays. For example, interrupts will stretch out time for anything using software delay loops since the loop will stop incrementing during the interrupt routine.

[2]The null statement is not necessary within the braces (at least for the Franklin C compiler).

Figure 3–9a Delay by Software Looping—C

```
1  public msec
   msecm segment code
   rseg msecm
    msec: jz x ;quit if a = 0
2       mov r0,#250 ;250 x 4 = 1000
3   z: nop
       nop
       djnz r0,z  ;4 uSec per loop
       djnz acc,msec
    x: ret
   end
```

[1]This module is made public so it could be linked with a program needing a time delay (such as Fig. 2-3b in Chapter 2). The details are discussed in Chapter 6.

[2]The delay could be fine tuned to compensate for the overhead at the start and the testing of the accumulator for the end. Technically the delay is still not precise, since there are 4 microseconds lost for the l c a l l and r e t.

[3]The n o p s bring the inner loop up to 4 microseconds so the index (r 0) can fit in a byte for a 1 mSec delay. Note that this routine destroys the value in r 0.

Figure 3–9b Delay by Software Looping—ASM51

```
   MSECM:DO;
1  MSEC:PROCEDURE (X) PUBLIC;
   DECLARE X WORD MAIN;
   DO WHILE X>0; /*# OF MSECS*/
2    CALL TIME(10); /*10 x 100 = 1000*/
     X=X-1;
   END;
   END MSEC;
   END MSECM;
```

[1]The procedure is made public so it can link with some other module needing a time delay (as the example in Fig. 2–3c). Unlike the very brief assembly example in Fig. 3–9a, this sets X as a WORD to allow longer delays.

[2]PL/M has a built-in TIME function which gives 100 microsecond delay for each value passed (assuming a 12MHz crystal). The value passed can only be a byte in PL/M 51, so the maximum delay can only be 25.6 mSec.

Figure 3–9c Delay by Software Looping—PL/M

beyond this is difficult, but all 3-byte instructions take 2 microseconds, all branching instructions take 2 microseconds, all 1- or 2-byte logical and arithmetic instructions take 1 microsecond, and the register-to-register moves take 2 microseconds. In summary, your best bet is to look it up! The example that follows is a delay routine, which takes about 1 mSec for each unit passed to it. The example using time delay passes a value of 50 to get a delay of about $50 \times 100 = 5000$uSec = 5mSec. Routines and passing parameters are discussed in detail in Chapter 5—the focus here is on the loops involved.

The obtaining of delay by software looping is quite inefficient when other processing could be going on. Software delays make the controller blind to other inputs while in the delay loop unless interrupts are added. This is discussed much more in Chapters 8 and following.

That completes the block constructs. Chapter 4 elaborates on pointers and based variables, which leads into Chapter 5, where subroutines, procedures, and functions—the heart of modern programming—are discussed.

REVIEW AND BEYOND

1. Show how a program can exit an iterative DO or for block early without using a goto.
2. What are possible pitfalls of unstructured languages?
3. It is argued that the goto is unnecessary in a structured language. What situations would possibly be easier to handle with a goto than with other structures?
4. What are the restrictions on use of goto in the two languages?
5. What is the difference between a while and a do...while in C? Which does PL/M support, and how could the other be "fabricated" using the other structures?
6. How do a DO CASE and a switch operation differ? What happens in C if you omit the break?

Chapter 4

Pointers
and Based Variables

STRENGTH OF C (AND PL/M)

Along with the support of functions (subroutines/procedures) in Chapter 5, the ability to reference variables by pointers is one of the strengths of C. This chapter starts with arrays, goes on to structures, and then plunges into pointers, based arrays, based structures, unions, and a few of the many obscure combinations of the whole lot. Just-in-time learning suggests that you only study as far in as you see a need now, and then just skim the rest for future reference. These types are best learned with a specific need at hand.

ARRAYS

An array is a collection of the same type variables referenced by a common name. There is a close relationship between arrays and pointers, which are discussed next, but in their simplest form, arrays do not require you to understand pointers. For all three languages, arrays involve storage in a block of *connected* memory. For a c h a r array, the successive bytes would occupy successive memory locations. For a i n t or WORD array, they would occupy successive byte pairs. Long and f l o a t arrays would have groups of four bytes, and, depending on the compiler, pointer arrays would have 1-, 2-, or 3-byte groups. In assembly you would use a group of locations by storing a *starting address* for the array and then adding to it the value of the *index*. Supposing the fourth element of the array is desired, the start address is fetched and the 3 is added to it. Incidentally, that is why both C and PL/M begin counting array elements with 0—the index can be directly added to the start address. The first element, having index 0, is the start address with a zero added. Unlike humans, computers prefer to start counting with zero rather than one. For assembly language arrays using *internal* memory, there are powerful instructions using r0 and r1 as pointers. It is, however, much more common to put arrays in off-chip RAM or ROM where there is more available memory. Figure 4–1b shows a subroutine that returns the r2th element of an array in off-chip RAM whose starting address was loaded into dptr. It adds the value in r2 to the pointer to set the pointer to the correct place in the array.

> STARTING ARRAYS WITH 1 RATHER THAN ZERO: The "one-th" element is the second byte.

> LETTING THE ARRAY INDEX REACH THE ARRAY SIZE: The index for a 20-byte array should never go above 19 since that is the top element. There is no error checking if you overrun the array; the variable that happens to be stored at the next higher address will unintentionally be involved.

Although the same sorts of instructions result from high-level languages, arrays are much easier to visualize in high-level languages. Figures 4–1a and 4–1c show the setting up of an array, ary, that has 20 two-byte members, and then another line copying some variable, x, into the tenth place. In Fig. 4–1a, the compiler assigns the values to the eighteenth and nineteenth locations above the start of the array because they are 2-byte values.

PL/M has only one-dimension arrays. There is a possibility of an array of structures where the structure happens to be an array— that is covered in a later section. In C there can be at least two-dimensional arrays. Some compilers sup-

```
unsigned int ary[20];
unsigned int x;
ary[9] = x;
```

Figure 4–1a Accessing an Array Element—C

```
fetch: mov a,r2   ;get the index
  add a,dpl     ;dpl is at 82h
  mov dpl,a
  clr a
  addc a,dph    ;dph is at 83h
  mov dph,a
  movx a,@dptr  ;get the array value into a
  ret
```

Figure 4–1b   Accessing an Array Element—ASM51

```
DECLARE ARY(20) WORD;
DECLARE X WORD;
ARY(9) = X;
```

Figure 4–1c   Accessing an Array Element—PL/M

LARGE ARRAYS: Be careful to size arrays only for what is needed. When most of an array is not used—especially with a multidimensional array—you can waste large amounts of memory. Embedded controllers, perhaps unlike large multi-use systems where the memory blocks are large and are "there anyway," should not occupy unnecessary RAM. Memory is not just sitting around free for use in an embedded system; if called for in the software, it will have to be added to the project boards.

port more dimensions, but the use is quite limited because it is very easy to run out of memory that way. A 10×10×10 array of floating values requires almost 4K bytes. A 25×25×25 array exceeds the entire possible 64K byte address space. Figure 4–2 shows a two-dimensional array of floating-point variables with a particular value being retrieved and put into another floating-point variable. (After the discussion of pointers, which follows, a second way to use arrays in C will be discussed, but this is the more obvious way for beginners.) Note that in this case ([5][0]) the compiler points to the 51st element from the start of the array (C8) because the first of the "rows" of 10 elements each take up 40 bytes.

A second example of a two-dimensional array shows the initialization for a series of strings as messages.

```
float xdata ary2d [10][10];
float xdata x;

x = ary2d[5][0];
```

Figure 4–2a   Two-dimensional Array—C

```
1   uchar code msg[][17]=
2   {{"This is a test",\n},
     {"message 1",\n},
     {"message 2",\n}};
```

[1]This is a two-dimensional array. The second size must be entered since it is not determined from the data list. The first dimension (3) is determined from the constant list.
[2]Two-dimensional arrays need *two* sets of curly braces for initialization.

Figure 4–2b   Two-dimensional Array—C

LOOK-UP TABLES

The use of arrays is well-suited to look-up tables as shown in Fig. 4–3a. In many embedded applications it is preferable to use tables rather than mathematical computations because a look-up can be executed quite quickly and usually involves less code. The tables can be computed in advance and included in the ROM space. It is possible to interpolate between points in a table to increase the accuracy of the answer or decrease the size of the table. Many science-oriented users, brought up on equations, expect a controller to carry out the math on demand to a high precision, which is probably totally unnecessary for most applications.

CONSTANT ARRAYS: It is often much quicker to look up values in a table than it is to compute them. The 8051 has an instruction specifically for fetching values from a table. Even for precision above 8 bits it is often sufficient to linearly interpolate between table values and can involve less code space than a complex floating point algorithm.

```
        #define uchar unsigned char
1       uchar code temptbl [] = {32, 34, 36, 37, 39, 41};
2       uchar x;


        uchar ftoc(uchar degc) {
3         return temptbl [degc];
        }
        main () {
          x=ftoc (5);
        }
```

[1]In this case the arrays are set up to be in the EPROM space and are made *unsized* in the sense that the size is determined by the number of entries in the list. That helps avoid the effort and error possibility of counting the list to see how many there are. Also, the define of "uchar" is much simpler to type and allows the attention to focus on the variable rather than on the word "unsigned"—the type that is preferred for embedded applications.

[2]Although the examples specify that the array be located in code or xdata space, realize that that is not *standard*. The initialization of an array would normally involve an entire series of program instructions *automatically* generated to move the code to RAM space at startup. Standard C expects that all variables not otherwise initialized will be set to zero. In embedded 8051 systems that is nonsense! Good compilers should make it possible to suppress that unnecessary and wasteful initialization.

[3]The returning of a parameter is discussed in Chapter 5, where functions and procedures are formally introduced.

Figure 4–3a   Temperature Conversion—C

```
        ;works with range of 0-5 degrees C input
        ftoc: mov dptr,#temptbl ;point to the table
1       movc a,@a+dptr  ;get it from the table
        ret

        temptbl: db 32,34,36,37,39,41
```

[1]In assembly, there is a handy instruction for accessing arrays in ROM space (often used for fast look-up conversion tables), the movc command. Since a is added to dptr, a subroutine to convert degrees Celsius to degrees Fahrenheit could work quite simply. Since the movc instruction includes the addition step, the operation is quite efficient. Many years ago, the initial developers of the 8051 obviously had this sort of application in mind.

Figure 4–3b   Temperature Conversion—ASM51

```
X:DO;
    DECLARE TEMP BYTE;
1   DECLARE TEMPTBL(*)BYTE
            CONSTANT(32,34,36,37,39,41);

    FTOC:PROCEDURE (DEGC) BYTE;
     DECLARE DEGC BYTE;
     RETURN TEMPTBL(DEGC);
    END FTOC;

    TEMP=FTOC(36);
END X;
```

[1]In this case, the arrays are set up to be in the EPROM space and are made *unsized* in the sense that the size is determined by the number of entries in the list. That helps avoid the effort and error possibility of counting the list to see how many there are.

Figure 4–3c Temperature Conversion—PL/M

STRUCTURES

A **structure** is a group of related variables referenced by one name. For assembly language, structures are handled only indirectly by keeping track of an index. In C and PL/M, however, structures can be very useful. It may help to think of a two-byte variable as a structure made up of two bytes. The first byte is the high part of the binary number, and the second byte is the low part. When you make an array of 2-byte values, each one consists of a high-low byte pair. Together they make up the entire number.

On a larger scale, a structure could represent the information about the solenoids of the sequencer in Chapter 1. One part could be the 32 bits that represent the 32 solenoids. A 1 could mean that a solenoid is "on." Then another part of the struc-

CONFUSING A STRUCTURE ELEMENT WITH A VARIABLE BY THE SAME NAME: Forgetting the overall structure name and dot when referencing a structure element may produce no compiler error if there happens to be *another* variable with the same name as the structure element or with the original (K&R) C it will assume a new variable of type int with a warning to you.

ture could be the amount of time the state should be held. A third part could represent whether the particular state is the last in the sequence. A DECLARE or define for such a structure is shown in Fig. 4–4. See also the solenoid example of Chapters 9 and 10. Again, assembly is not shown because the treatment is up to the programmer to include with index pointers and starting addresses to access a (contiguous) block of bytes.

```
      struct {
         unsigned long s;
 2       unsigned int t;
         unsigned char done;
      } state;
 1    state.t = 321; /*use of structure element*/
```

---

[1]This sets up a single structure named **state** which contains seven bytes.

[2]The first four bytes are the 32 on/off bits for the solenoids, while the next two bits hold the 16-bit number which is the time to the *next* state transition. Finally, the last byte holds the 1 or 0 which indicates whether to go on to another state or start over.

Figure 4–4a   Structure Example—C

```
      1
DECLARE STATE      3    2
      STRUCTURE (S1 WORD,S2 WORD,T WORD,DONE BYTE);

STATE.T = 321; /*SAMPLE USE OF STRUCTURE*/
```

---

[1]This is a single structure named **STATE** which contains seven bytes.

[2]The first four bytes are the 32 on/off bits for the solenoids, while the next two bits hold the 16-bit number which is the time to the *next* state transition. Finally, the last byte holds the 1 or 0 which indicates whether to go on to another state or start over.

[3]PL/M cannot hold 32 bits as a single variable, so two 2-byte WORDs are used.

Figure 4–4b   Structure Example—PL/M

PUTTING BITS IN A STRUCTURE: Since bits can only be in one part of on-chip RAM, an error will result if you define a structure made up of both bits and other data types. A structure has to involve contiguous bytes.

STRUCTURE TEMPLATE

It is often the case that a particular structure *form* or **structure template** is used in several different structures. This can be done in both languages. In PL/M it is done by defining a LITERAL, whereas in C the language directly supports a named **structure tag**. This is shown in Fig. 4–5. Each language could use the name s t a t e f o r m repeatedly as new structures are set up.

```
#define uchar unsigned char
#define uint unsigned
struct stateform {
 unsigned long s;
 uint t;
 uchar done;
};

struct stateform state;
```

Figure 4–5a   Structure Template—C

```
DECLARE STATEFORM LITERALLY
    'S1 WORD,S2 WORD,T WORD,DONE BYTE';

DECLARE STATE STRUCTURE (STATEFORM);
```

Figure 4–5b   Structure Template—PL/M

ARRAY OF STRUCTURES

In the case of the envelope machine of Chapter 1 (as well as most any state machine), it would make more sense to use an **array of structures**. Taking the previous structure and making an array is quite direct, as shown in Fig. 4–6.

```
#define uchar unsigned char
#define uint unsigned
struct stateform {
 unsigned long s;
 uint t;
 uchar done;
};


struct stateform state [20];  4
      1    3     2
state[11].s |= 0x04000000;
```

[1]The array is handled with the [ ] of the usual array. The dot indicates that the particular element within the structure is coming next. Since C supports the 32-bit l o n g, the 32 b i t s can all be part of one element. The example involves 140 bytes of storage (20×7 bytes).

[2]C has to use hexadecimal notation. The hex representation looks long because it addresses the *entire* 32 bits. You should avoid decimal notation (as might have been done in BASIC) since it would obscure the fact that the variable is not being used as a number.

[3]Notice that the |= operation saves considerable space by not having to repeat the structure reference.

[4]Either the large memory model must be used or else the x d a t a keyword should come ahead of s t r u c t since this structure won't fit in on-chip RAM.

Figure 4–6a   Array of Structures—C

```
1  DECLARE STATE (20)STRUCTURE (STATEFORM) AUXILIARY;
2  STATE(11).S1=STATE(11).S1 OR 0000010000000000B;
   /*SETS A BIT TO TURN ON A SOLENOID*/
```

[1]The array is handled with the ( ) of the usual array. The dot indicates that the particular element within the structure is coming next. Thus, it is referencing one of the two 16-bit words. It now involves 140 bytes of storage (20×7 bytes).

[2]PL/M allows binary representation, so that was used to emphasize the *bit* nature of the variable. This addresses one of two 16-bit words. Whether you choose hexadecimal or binary notation, in no case should you use decimal notation (as might have been done in BASIC) since that would obscure the fact that the variable is not being used as a number.

Figure 4–6b   Array of Structures—PL/M

ARRAYS WITHIN STRUCTURES

Finally, it is possible to have **arrays within structures**. In the previous example it might well be more efficient in the final machine code to avoid reference to the larger variable types which, depending on the compiler, might bring in large, undesired libraries. The previous examples become those shown in Fig. 4–7.

```
#define uchar unsigned char
#define uint unsigned
struct stateform {
 uchar s[4];
 uint t;
 uchar done;
};


struct stateform state [20];

1  state[11].s[0] |= 0x04;
```

Figure 4–7a   Array of Structures with Arrays—C

```
DECLARE STATE (20)
STRUCTURE (S(4)BYTE,T WORD,DONE BYTE)
                       AUXILIARY;
1  STATE(11).S(0)=STATE(11).S(0) OR 00000100B;
   /*SETS A BIT TO TURN ON A SOLENOID*/
```

ᴵThis is the 78th byte of the structure.

Figure 4–7b   Array of Structures with Arrays—PL/M

This is the limit for PL/M and was hinted at earlier as a way to get the effect of two-dimensional array—an array of structures with the structure being another array. The C language can go on from here with **nested structures**. It is a challenge to come up with a real embedded control example, but I suppose you could take the previous example and add a separate piece of information related to incoming sensor pulses for repair logging. It could then look like Fig. 4–8.

This is not the height of simplicity and is probably unnecessary for embedded applications, but it does see use for data processing where a structure might hold an employee's name and address and other structures could *use* that structure template with additional parts for other information. If you would *really* lose the casual reader, nest it all in one expression without tags! In addition, as Fig. 4–9 shows, put it all on as few lines as possible. It won't bother the compiler, because line feeds are ignored, but it will certainly amaze your friends!

```
#define uchar unsigned char
#define uint unsigned
struct stateform {
 uchar s[4];
 uint t;
 uchar done;
};

struct repairtype {
 struct stateform state;
 uchar sensorcount;
};

struct repairtype repairs [20];
```

1   `repairs[11].state.s[0] |= 0×04;`

___

[1]This is the 89th byte of the structure (I didn't count but relied on the results of a simulator to tell this!).

Figure 4–8   Nested Structures—C

___

```
#define uchar unsigned char
#define uint unsigned
```
1
```
struct {
 struct {
  uchar s[4];
  uint t;
  uchar done;
  } state;
 uchar sensorcount;
 } repairs [20];

repairs[11].state.s[0] |= 0×04;


struct{struct{uchar s[4];uint t;uchar done;} state;
 uchar sensorcount;} repairs [20];

repairs[11].state.s[0] |= 0×04;
```

___

[1]The best indentation style for this is unclear. I would tend toward packing more on each line, but sometimes it is desirable to fit comments in with each element of the structure.

Figure 4–9   Nested Structures without Tags—C

MEMORY SPACES AND EFFICIENCY

With the 8051 family of controllers, as was discussed in Chapter 2, there are the 3 or 4 different *types* of memory. In assembly you make the choice quite deliberately— there are only a few instructions that will address the off-chip memory. You might review the movx and movc instructions in Appendix A to see this. In high-level languages, you make the decision when you DECLARE or define the variables. In PL/M, the words AUXILIARY, CONSTANT, or the (default) MAIN determine where the compiler will keep variables. In 8051 C compilers some sort of extensions like xdata, code, or data determine the type of memory used. In the 8051 C compilers, the choice of **memory model** determines the default memory type used. If small is used, the compiler will store every variable *not otherwise specified* in on-chip memory. For small programs with few variables, there is enough memory space. Larger program models by default assign variables to off-chip memory.

There are some C compilers that will allow **bank switched** memory schemes when the 64K byte limit is reached. By having the compiler automatically add instructions to toggle hardware bits that enable one or another bank of RAM chips, it is possible to have several blocks of RAM or EPROM much like the 8086 families have segments. This is very unusual because programs *that* large probably ought to use a different sort of computer! Bank switching might make sense where there is already extensive software development and it is *then* discovered that the project was growing too large to fit. Then, in order to save all the expensive development effort, bank switching might be attractive.

In the same category as bank switching falls the **stack-oriented** approach to putting C on the 8051. Because the 8051's hardware stack is in internal on-chip RAM, which is very small, and because *standard* C is highly stack-oriented, some compilers make a stack in off-chip RAM using software instructions. The call return addresses are put on the internal stack, but all the other passed parameters and perhaps even temporary results go on this artificial stack. Even a cursory knowledge of the instruction set and architecture will tell you how inefficient this can become. This entire area of fitting C to the 8051 makes the choice of compiler for the individual application quite important. If the effort is to fit huge code into this small microcontroller, an *earlier* question should have been, *"Does the application fit into one 8051?"* There are other processors that do large computing jobs much more efficiently, and there are methods of incorporating several processors or controllers in one project, as discussed in Chapter 12.

ADDRESSES

To move to the area of really powerful programming, it is necessary to think some more about the memory of a computer. Starting in Chapter 2, we have been talking about variables by their *name* and leaving the details of *where* to store them for the compiler or assembler. Even in assembly, the use of the ds directive and relocatable segments can allow the exact location where the value of a variable is kept to be determined at the time the pieces are linked. Chapter 6 has more details on linking and modular programming, but all three languages easily leave those decisions up to the linker unless you *purposely* override them. With embedded systems, you need specific addresses for I/O devices; there is usually no prewritten driver to insulate you from such details. Port chips, A-D chips, and other added I/O have absolute addresses determined by the wiring and address decoding of *your* design. If the compiler/linker system chose to put the I/O at some other address, it would cause disaster. In the general scheme of things, though, one memory location is as good as the next so long as the routines and modules are all consistent in going to the same place for the same variable.

Chapter 5 will discuss routines (subroutines, procedures, or functions depending on the language involved). At that point it will be shown that routines are most powerful if the variables to be processed can be *passed to* the routine. In assembly you would load up some registers, and in PL/M or C you would put the variables in parentheses when the routine is called. This works well when only a few values need to be passed and when no more than one value is returned. But imagine you want to send messages to a display and these messages could be 3 characters or 80 characters long, but you want *one* routine to handle them all. Or suppose you have a routine that works through a list of data just in from an A-D converter to smooth out data that might seem out of line due to noise. You want this routine to work on the data "where it sits" rather than taking extra processor time to move it somewhere else. In either case if you can just send the *address* of the data rather than the entire list of data, the routine can go to that address and get the data only as needed. At different times the same routine can work on data at different addresses. It is this desire to make routines more flexible and efficient that really makes pointers and based variables important.

POINTERS

A **pointer** is a *variable* that holds the *address* of another variable. In assembly it could be a byte put into r0 or r1 for the internal memory space, or else a 2-byte value put into dptr for access to external RAM or code (EPROM) space. A high-level language example of setting up a pointer to a variable and then using the variable pointed *to* is shown in Fig. 4–10.

UNIVERSAL POINTERS

The PL/M pointer is *always* a WORD even if the based variable is defined in on-chip RAM (needing only an 8-bit address). In C for the 8051 there is a significant variation between compilers. Some have *universal* pointers by including a third byte that holds

```
  #define uchar unsigned char
  uchar count;
1 uchar *x;
2 uchar xdata *y;
  uchar data *z;
  uchar code *w;
  uchar data *xdata zz;

3 x = &count;
4 *x = 0xfe;
```

[1]The pointer is not defined separately. Rather, it is defined in the process of naming the based variable. The example says there is a byte variable found where x is pointing.

[2]This line and the following three show the compiler-specific keywords of the Franklin compiler. The first three are specifying where the variable pointed *to* is located. The pointer will not be a universal one as discussed in the next section. The last definition line sets up a pointer, z z, specifically located in off-chip RAM pointing to a variable located in on-chip RAM. Where the *pointer* is located is established by the choice of the memory model in the compiler invocation or the pragma at the top of the file. It is now possible to even mix modules compiled under different models with the techniques of Chapter 6.

[3]Usually a routine for processing an array will be passed the *address* of the array. This is done where it is called by using the & operator which returns *the address of* the variable that follows.

[4]This line actually puts the constant in the place *pointed to* by x.

Figure 4–10a Based/pointer Variable—C

```
  DECLARE COUNT BYTE;
1 DECLARE XPTR WORD;
  DECLARE (X BASED XPTR) BYTE;

2 XPTR=.COUNT;
  X=0FEH;
```

[1]The names are not significant—X could be based on Y if you preferred. Also, X does not have to be a BYTE; it could be a WORD or, as will be discussed later, an array or a structure. The only definite requirement is that the base, XPTR, must be a WORD.

[2]Usually a routine for processing an array will be passed the *address* of the array. This is done where it is called by using the dot operator. The **dot operator** gets *the address of* the variable that

Figure 4–10b Based/pointer Variable—PL/M

a code identifying which type of memory space is involved. (The codes for Franklin's pointers are 1=idata, 2=xdata, 3=pdata, 4=data, and 5=code.) For the different address spaces, other compilers use only two-byte pointers (like PL/M) or one- and two-byte pointers. The three-byte pointers allow library routines to accept pointers to *any* memory space and then internally decide which set of code to use for processing. Otherwise, the programmer or the compiler must know which memory space can be used and carefully use the right space with the right library routine. In all of these situations the compiler can affect code efficiency. Friendliness to the programmer is useless if the application can't get done in time, but efficiency is no great benefit if the processor has very little to do and no significant time constraints. This will be discussed further in Chapters 8 through 12 on real-time applications.

UNINITIALIZED BASE: If you use a based variable *before* you set the pointer to a specific place, it will either point randomly (PL/M) or point to zero (C). The effect could be bad in either case. In Fig. 4–10, a 254 is put into the variable named `count`, but if the value for x (or XPTR) were *uninitialized*, it is like randomly firing a loaded gun; if you pull the trigger you might hit anything!

ADDRESSING THE WRONG MEMORY SPACE: By mixing up `code`/`CONSTANT` and `xdata`/`AUXILIARY` pointers, when you send a message to a display routine, if the message is a "canned" one in EPROM, it could be brought from xdata space (the RD line on the 8051) rather than code space (the PSEN line on the 8051). If the display routine works out of code space only, sending it the *address* of a RAM message will have it fetch values from the *same* numeric address in the code space—certainly the wrong data.

ARRAY POINTERS

Things rapidly become complicated when you move with pointers to arrays. PL/M is fairly restrictive, but C can go wild! A *based* array is straightforward enough, as shown in Fig. 4–11.

WRONG POINTER INCREMENTING: If you are using a pointer to move through an array of 2-byte variables, be careful! Adding 1 to a pointer moves the address by one (I think!). But, with C, it may be that incrementing with a ++ may move it by two, since it knows the pointer is to 2-byte quantities. In PL/M it is up to you to know the size and handle the pointer arithmetic yourself, but with C there are varying degrees of "helpfulness" that may get you into trouble. Array indexing, of course, moves the correct number of bytes for the data type.

```
     #define uint unsigned
1    uint xdata a[];
     a[22] = 0xff;

     uint xdata *a;
2    *(a+22) = 0xff;
```

[1]The two examples are *exactly equivalent*, and the preference depends on the notation you want to use in the routine. On large machines it is said that the latter approach compiles to faster code, but that may not be the case with the 8051.

[2]Notice that incrementing by 22, *(a+22), will add *44* to the value in a since a points to (*two*-byte) integers. Thus, the compiler will determine the type of variable involved and increment by the number of bytes per element.

Figure 4–11a Based Arrays—C

```
     DECLARE APTR WORD AUXILIARY;
     DECLARE (A BASED APTR) (100) WORD AUXILIARY;
1    APTR=0FA7H;
     A(22) = 0FFH;

     DECLARE APTR WORD AUXILIARY;
     DECLARE (A BASED APTR) WORD AUXILIARY;
1    APTR=0FA7H;
2    APTR = APTR+44;
     A = 0FFH;
```

[1]The two examples are different in that the first treats A as an array while the second treats it as a single element and moves the pointer.

[2]Notice that where the array is incremented by 22, A(22), *44* is added to the starting location of A since it holds *two*-byte values. The pointer can be moved manually, but it is then up to the programmer when using based variables to recognize that WORDs take two bytes.

Figure 4–11b Based Arrays—PL/M

ARRAYS OF ARRAY POINTERS

Particularly confusing to programmers who started on some other high-level language is the relationship in C between pointers and arrays. The name of an array *is* a pointer. To add to the confusion, you can have a pointer to an array and you can have an array of pointers! You can even have a pointer to an array of pointers (called **multiple indirection**). In the example of Fig. 4–12, an array of pointers is used for a group of display message *pieces* that are strung together when sent to the display device. The example keeps all the pieces in code space as well as the pointer array, but if some pieces were canned headers whereas other pieces were current values of variables (in RAM), it would be necessary to either use the generic (3-byte) pointers of some C compilers or else make an array of structures

```
     #include <reg51.h>
     #define uchar unsigned char
     uchar code m1[ ]={"this is a test"};
     uchar code m2[ ]={"you failed"};
     uchar code m3[ ]={"you passed"};
     uchar code m4[ ]={0};
1    uchar code *code fail[ ]={&m1[0],&m2[0],0};
     uchar code *code pass[ ]={&m1[0],&m3[0],0};

     void display(uchar code **message){
      uchar code *m;
2     for (;*message!=0;message++){
3      for (m=*message;*m!=0;m++){
        P1=*m;
       }
      }
     }

     main(){
      display(&pass[0]);
     }
```

¹The message-piece *pointers* are collected together into arrays which are also put in code space. This is particularly useful if the RAM is only on-chip, because the pointer array then requires none of that very limited resource.

²This line walks through the array of pointers until a zero pointer is found. Since the initial value of message is already set, no initialization is needed in the f o r loop.

³This line sets m to point to the start of one of the message pieces and then increments m through the piece until a null is found indicating the end of that message piece. The code definition with "s automatically adds a null (0) at the end of the string.

Figure 4–12a   Multiple Indirection—C

```
     DECLARE AS LITERALLY 'LITERALLY, EOM AS '0';
     DECLARE M1(*) BYTE CON ('THIS IS A TEST',EOM);
     DECLARE M2(*) BYTE CON ('YOU FAILED',EOM);
     DECLARE M3(*) BYTE CON ('YOU PASSED',EOM);
     DECLARE M4(*) BYTE CON (0);
1    DECLARE FAIL(*) WORD CON (.M1(0),.M2(0),EOM);
     DECLARE PASS(*) WORD CON (.M1(0),.M3(0),EOM);


     DISPLAY:PROCEDURE (X);
     DECLARE (X,Y) WORD;
     DECLARE (MESSAGE BASED X) WORD CONSTANT;
2    DECLARE (M BASED Y) BYTE CONSTANT;
3    DO WHILE MESSAGE<>0;
      Y=MESSAGE;  /*ASSIGN BASE FROM LIST*/
      DO WHILE M<>0;
       P1=M;
       Y=Y+1; /*MOVE TO NEXT CHARACTER*/
      END;
      X=X+2; /*POINTERS ARE 2 BYTES ! */
     END;
     END DISPLAY;


     DO WHILE 1;
      CALL DISPLAY(.PASS(0));
     END;
     END Z;
```

[1]The message-piece *pointers* are collected together into arrays which are also put in constant space. This is particularly useful if the RAM is only on-chip because the pointer array then uses none of that very limited resource.

[2]There is no support of two-dimensional arrays directly—hence the need to fetch the pointer out of the first array and assign it to the pointer of the second array.

[3]No space is actually allocated for the array. The easiest way to grasp this is to see it as a series of operations on an address. First the pointer to the pointer array X is brought in. Then the pointer found at *that* location is transferred to the second pointer, Y, so it can be incremented through the message piece. If you preferred, the M variable could be made a based *array* which could be in code space, but it is still illegal to have a based variable serve directly as a base for another variable.

Figure 4–12b   Multiple Indirection—PL/M

that have one byte designating the particular memory type and a second two bytes that are the actual address pointers. With the pointers to the message *pieces* in an array, the pointer to the pointer array can be passed to a display routine, which will work through the entire string. Although this looks quite complicated, if you are pressed to save code space for messages, you can go as far as identifying repeated phrases and reusing the code with this sort of pointer system.

Notice that each array referenced by the array of pointers does not need to be the same size. This is one form of what is sometimes called a **sparse array**. Another form, more common in data processing on large computers, is the linked list. For the example here, the compiler trusts that indices will be kept in bounds by some other means. Usually in message arrays, either the length of the message is specified as the first byte of the array, or else a special character is put at the end—usually a nonprinting character that would not otherwise occur in a message. C automatically inserts a null, /0, at the end of any defined string, but PL/M requires the addition of an "end of message" by the user. For predefined arrays, PL/M has the SIZE operator whereas C has various library functions such as sizeof and strlen to do similar things. Check the language compiler manual for more specific details on those functions. It is probably equally efficient to use an end-marker or an initial-length number, but the former is more flexible if you revise your messages or string them together.

STRUCTURE POINTERS

It is a very small step from based arrays to based structures. Having left off the discussion with arrays of structures holding arrays, a real application for *based* structures is the message format for DCX/DCE (Chapters 8 through 12, also shown here in Figure 4–13). One task communicates with another task by "sending" a message. This sending involves giving the message pointer to the operating system, which passes the message or the pointer to the receiving task. Both tasks agree on the structure of the message, but the actual storage location is not fixed in the system.

```
#define uint unsigned
#define uchar unsigned char
struct msg1 {uint lnk; uchar len,flg,nod,sdt,cmd,stuff;};
struct msg1 *msg;

void rqsendmessage(struct msg1 *m);

main(){
  uchar stuff;
  msg->len=8;
  msg->flg=0;
  msg->nod=0;
  msg->sdt=0x12;
  msg->cmd=0;
  msg->stuff=stuff;
  rqsendmessage(msg);
}
```

Figure 4–13a   Based/pointer Structures—C

```
X:DO;
$INCLUDE (REG51.DCL)
DECLARE PTR WORD MAIN, STUFF BYTE;
DECLARE (MSG BASED PTR)
  STRUCTURE (LNK WORD,
  (LEN, FLG, NOD, SDT, CMD, STUFF) BYTE);

RQSENDMESSAGE:PROCEDURE EXTERNAL;
END RQSENDMESSAGE;

MSG.LEN=8;
MSG.FLG=0;
MSG.NOD=0;
MSG.SDT=12H;
MSG.CMD=0;
MSG.STUFF=STUFF;

DPH=HIGH(PTR);
DPL=LOW(PTR);
CALL RQSENDMESSAGE;
END X;
```

Figure 4–13b   Based/pointer Structures—PL/M

UNIONS

Usually included with structures is another data type called a **union**. A union is, as the name implies, a union of different data types; the difference from a structure is that the union uses different names and types for the *same* space. The most obvious example, shown in Fig. 4–14, is where you want to store a 16-bit timer value which is read in as two bytes. While you could use a cast to an integer and an 8-bit shift to get the high bits in place, it is also possible to define a union made up of a 2-byte structure and an integer. When you want to fill the high byte, refer to the bytes, but when you want to use the result, refer to the integer.

Going one step further, a union is very useful for message structures which are assigned by an operating system as in Chapter 10. When a message arrives for a task, the receiving task must understand the message by a pre-arranged pattern. If several different forms are possible depending on where the message comes from, a union makes it easier to lay a different "template" over the sequence of bytes depending on the source. The example of Fig. 4–15 shows such a union for a DCX/DCE message. Incidentally, when you say you are using a based structure made up of a structure and a union of structures, you will again impress your

```
#define uint unsigned
#define uchar unsigned char
1   union split{uint word;
                struct{uchar hi;uchar low;}bytes}};
    union split newcount;


2   newcount.bytes.hi=TH1;
    newcount.bytes.lo=TL1;
    oldcount=newcount.word;
```

[1]The first line sets up the *tag* for a union type called "split." The parts are "word" and "bytes"—the latter consisting of a structure with two bytes called "hi" and "lo." The only part that becomes implementation specific is the decision that "hi" comes before "lo," which is the case for the normal 8051 arrangement of bytes. This might not transfer to another processor.

[2]The reference to the structure elements has three parts, all separated by the dot (.).

Figure 4–14   Union of Integer and Bytes—C

```
#define uint unsigned
#define uchar unsigned char
1   union mtag{struct {uint keycode;}kmsg;
               struct {uchar cursor;uchar dat[12];}dmsg;}

2   struct msgform {struct header hdr;union mtag dat;}*msg;

    keystring[i]=(msg->dat.kmsg.keycode & 1)+'0';
3   msg->dat.kmsg.keycode>>=1;
    msg->hdr.cmd=0×40;
    msg->dat.dmsg.cursor=0;
    for(i=0;i<=10;i++){
4     msg->dat.dmsg.dat[i]=keystring[i];
    }
```

[1]This is the definition of the tag for a type of union, here called "mtag." It consists of two structures named "kmsg" and "dmsg" for the two anticipated data arrangements which depend on the message source. The two structures are of different sizes. That is immaterial since the whole thing is to be based, but, were it fixed, the union would be allocated the larger of the two sizes.

[2]This line actually defines the message (which is *based*). The use of the tag name simplifies this line, but it would be legal to substitute the two structures if you preferred. Taking it step-by-step with tags may help the beginning programmer and won't hurt even experienced programmers. The header is another structure (not defined here) which holds the message address, length, and other related information common to all messages.

[3]The use of a based structure requires the -> for the first separator, but you can see all the dots involved in reaching into the union (dat), into the specific part of the union (kmsg), and to the specific part of the structure (keycode).

[4]This shows reaching into an array within the structure.

Figure 4–15   Union of Structures—C

friends who claim to "know" C! If you mix in arrays of pointers to structures made up of arrays, you may even amaze yourself, but it *can* be disentangled if you understand the basics.

REVIEW AND BEYOND

1. How many bytes would be set aside for a 10-element WORD/int array? Would the low-order bytes be in a group and then the high-order bytes, or would they be byte-pairs? If the array started at location 2020h, where would the 2 bytes of A(5) or a[5] be found?

2. Can you have two-dimensional arrays in PL/M? Explain.

3. With the 8051, why are arrays of dimension greater than 2 quite uncommon?

4. Set up a structure to hold coordinate values (say for drawing graphs in x-y space).

5. Set up and initialize two arrays of 2-element structures holding x-y coordinates for drawing, say, a square. Then outline a plotting function with a *based* equivalent to use inside the function. Show how you would call the function to plot a square.

6. What are the different memory spaces in the 8051? Can the same address apply to different spaces?

7. What are the possible solutions to the problem of having different memory spaces when pointers are used? What are some of the trade-offs involved?

8. Why is the printf() function of C more complicated with the 8051 than in normal "flat" address space computers?

9. Explain the difference between pointers to arrays and arrays of pointers. Give an example of each.

10. Is there any difference between an array and a pointer?

11. Write out examples of a structure and a *based* structure including the way to reference a member of the structure.

12. What are some purposes for unions?

Chapter 5

Routines

SUBROUTINES, PROCEDURES, AND FUNCTIONS

The terms subroutines, procedures, and functions are used by the three languages to refer to the same thing. From here on, for simplicity I'll use the term **routine** since it suggests something that is repeatedly done the same way. Because programmers are always looking for short cuts, finding the same code in several places leads to the thought, *"I shouldn't have to write this code over and over."* That is where routines come into play. A routine can do such things as producing steps to drive a stepper motor, or converting a number to a displayable (ASCII) form. The details can be "packaged" in one place. The main program can **call** the routine when needed. When the routine finishes, it issues a **return**, and the main program flow resumes with the next instruction. The same routine can be called elsewhere, and the code of the routine can be used over again. An example is the routine to drive a stepper motor in Fig. 5–1.

Figure 5–2 shows a part of a main program using the routine of Fig. 5–1 to take 20 steps of a stepper motor. The stepper motor used here is a standard *unipolar* four-phase motor. Unipolar, as shown in the schematic, can be driven with four transistors—one to ground each of the four windings. The alternative, *bipolar* steppers, is more efficient of copper having only two windings, but the drivers have to source as well as sink current so an "H" bridge arrangement is necessary. Such drivers are available in ICs, but they are not as readily available or inexpensive as simple transistors. Also, it is possible to obtain driver chips that translate to the phase pattern from two lines— one for step and one for direction. Such logic, involving an up-down counter and some gates, is an excellent example for programmable logic devices (PLDs).

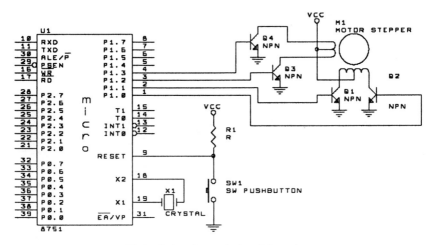

Figure 5–1a   Stepper Driver Schematic

```
#include <reg51.h>
uchar code pattern[ ]={0x5,0x9,0xa,0x6};

1   void step(void){
2     static uchar i;
3     P1=pattern[i=++i&3];
4     msec(8);
    }
```

[1]With the advent of ANSI standard C, the use of the word v o i d is encouraged when nothing is passed *to* or returned *from* a **C function**. Passing parameters is discussed later in this chapter.

[2]The index is made s t a t i c to be sure it doesn't move between steps. The motor needs the transitions to line up with the table if the motor is to move smoothly.

[3]Here is a typical "condensed" line in C. Because of the precedence (see end of Chapter 2), the ++ precedes the & and the assignment to i comes before the actual value is used as the index for the array. The result is that each time the routine is used, the value of i goes up by one, but if it reaches 4, it is reset to zero.

[4]The delay routine was introduced at the end of Chapter 3. It is necessary to delay between phase transitions to allow the motor to catch up.

Figure 5–1b   Stepper Driver—C

```
    sptr segment data
    rseg sptr
    i: ds 1

    stepr segment code
    rseg stepr
1   step: mov a,i
2         anl a,#3
          mov i,a
          mov dptr,#pattern
          movc a,@a+dptr
          mov P1,a
          mov a,#8
3         lcall msec
          ret
4   pattern: db 0101b,1001b,1010b,0110b
    end
```

[1]An assembly language **subroutine** ought to have a label (here s t e p) so the call can be simplified. It is *possible* to call to an absolute address, but this would be difficult to manage as lines are added or for relocatable modules. The r e t marks the end of the subroutine and pops the return address off the stack to resume the program flow at the line just after the original call. It is not good to enter a subroutine by anything except a call. A jump into a routine would not push the return address on the stack so the return would do unexpected things.

[2]The index works around the pattern table, so it must not get above a value of 3. Rather than testing and resetting, since 4 is a nice round binary number, the masking does the resetting in one step and will keep things in bounds even the first time when the value of i can be anything.

[3]This routine was given at the end of Chapter 3.

[4]The pattern never changes so it is put in code space and would be ultimately burned in ROM with the program instructions.

Figure 5–1c   Stepper Driver—ASM51

EASE OF UNDERSTANDING

It is misleading to suggest that routines are just a reflection of the laziness of the programmer. In reality, as programs have become increasingly complex, it is *under-standability*, more than code savings, that has motivated the use of routines. Even for a routine used only *once*, it can be good to break it out of the in-line code. Naming a routine in such a way that the call describes the *function* of the routine gives you a quick general picture of what is going on without getting lost in the details. For example, a block of code to get the mean and standard deviation of a group of readings could be broken out as a routine and named p r o c e s s _ s t a t i s t i c s. You, as a new reader of the program, come to p r o c e s s _ s t a t i s t i c s in the main program. You don't have to understand all the details to appreciate what is to happen.

```
$INCLUDE(REG51.INC)
DECLARE I BYTE MAIN;
DECLARE PATTERN (*) CONSTANT=
              (0101B,1001B,1010B,0110B);

1   STEP:PROCEDURE;
2     I=(I+1)AND 3;
      P1=PATTERN(I);
      CALL TIME(80);
      END STEP;
```

[1]A PL/M **procedure** always has a name (here STEP) and the word PROCEDURE. It will always have an END and should have the label with the END. One additional feature, not shown here, is the ability to *pass parameters* to the procedure and to return a value. This is discussed in a later part of this chapter.

[2]This step moves the index through the pattern table while resetting it to zero after the end of the table is reached. The time routine comes automatically with PL/M and here gives a delay of 8mSec with a 12mHz clock.

Figure 5–1d   Stepper Driver—PL/M

```
for(i=1;i<21;i++){
1   step( );
    }
```

[1]The call is implicit with the naming of the function. Sometimes programmers add the word call with a #define call line at the top to allow the call to appear and improve understandability (ignored by the compiler). For certain 8051 family relatives with small on-chip ROM, the lcall and ljmp are not supported, and the compiler must be given the ROM(small) directive to avoid using them. The use of a function must be shown by the parentheses even though no parameters are being passed to the function.

Figure 5–2a   Calling the Routine—C

If you need more detail, you can go to the routine. But, if you are interested in the hardware interface to the A-D converter, obviously *this* is not the routine to study.

It has become a rule-of-thumb that no routine (or main module) should be longer than about **60 lines** or about a page. Some even argue for about 20 lines, which will fit on a normal CRT display. Anything that is longer than that ought to be broken into other routines so the reader doesn't have to follow a long run of in-line code. The ultimate disaster is a 20 page in-line program with jumps forward or back several pages. As discussed at the start of Chapter 2, the same rules should apply to flow charts as well.

```
      mov r7,#20
1   x:lcall step
      djnz r7,x
```

[1]Assembly has two kinds of calls. The l c a l l involves 3 bytes for the instruction and allows a full 16-bit location for the subroutine. The a c a l l is unusual. It fits the call into 2 bytes of instruction but only reaches over the 2k block of the next instruction. That is to say, the call address is made up of the top 5 bits of the current location combined with the bottom 11 bits found in the a c a l l instruction. While it is efficient for code space in small programs, it is a challenge for the assembler and linker programs when modules may be relocatable. Modules that use this are identified as "in-block" and are kept at specific 2K byte boundaries for the final addresses. With most assemblers, you can just use the word c a l l and the assembler will choose the form of call to use.

Figure 5–2b   Calling the Routine—ASM51

```
      DO I=1 TO 20;
1       CALL STEP;
      END;
```

[1]Parentheses are needed with the C A L L only if there are parameters to pass.

Figure 5–2c   Calling the Routine—PL/M

By picking a **descriptive name** for the routine, the *function* of the routine is apparent in the call. A routine helps keep the focus on the main purpose of the program, particularly when the details are very hardware-oriented. For example, look back at Fig. 5–2. It should be intuitively obvious what sorts of things are done in "step" and "delay."

Besides being more understandable to someone reading the program, routines do save code space. Because their code is in only one place and the *call* to the routine takes just two or three bytes, the savings with long, repeated routines can be considerable. The process of calling and returning *does* take a little more computer time than **in-line code,** but for a simple call it is only a few microseconds. If the routine involves more than one or two lines, the code savings and the increased understandability far outweigh the time cost of the call.

Routines reduce the amount of code and make programs easier to understand. They can also group all the hardware-specific details in one place. A **driver,** such as in Fig. 5–1, is a piece of software that serves as an interface between the rest of the software and some particular hardware (although sometimes the term applies to an interface to other software as well). As mentioned in Chapter 1 on task parameters, drivers put the machine-specific details in one place and are the *interface* between the hardware and software. Besides lumping the details in one place and *hiding* them from the rest of the program, drivers make it easy to adjust to hardware *changes*. Suppose the final PC board layout requires the switching around of a few bits on one port. If you have carefully used drivers, just changing a few lines of code in the drivers makes the software ready to go. You don't need to hunt through the entire code, finding all the places where the bits were used. Modular programming, to be discussed in Chapter 6, also helps in this situation.

Picking out prewritten drivers to match the hardware is called **configuring** the system. For large-computer systems, drivers come with the system software for many common pieces of hardware—printers, kinds of storage and memory devices, types of displays, and so on. *Embedded* systems (especially the 8051 family of processors) are usually small, are control-oriented, and have the computer as an integral part of the application hardware; there is no consistent, generally accepted set of hardware to attach. Because general-purpose drivers, where they exist, tend to be larger and less efficient than custom-written ones, you may be better off writing your own drivers.

NESTED ROUTINES

It is common to have routines make use of other routines. Saying they are **nested** means that one routine is used inside the other routine (like a bird in a nest?). To keep within the 60-line limit and to improve understandability as programs become more complex, it is not uncommon to call routines to a depth of 4 or 5. There are limits to nesting depth in some compilers, but such limits are not very restrictive. A more real limit with the 8051 family is the lack of stack space in internal memory. Each call puts 2 bytes (the return program-counter address) on the internal stack, so eventually there is no more internal stack space. On other processors, C compilers usually rely heavily on the stack for passing parameters, but the 8051 versions may have options that put all parameters on an artificial *external* stack. Five or ten levels of nesting without parameter passing is usually no problem even for on-chip stack, so, for small programs, it is generally safe to ignore the nesting and stack depth limits.

CONFUSION ON PASSING ARRAYS: If you want to pass an array to a function, give the name of the array without the brackets as in `display(array);`. If the parameter to be passed is a particular *element* of the array, give the name with a specific number within the brackets as in `step(phase[3]);`. If you want to pass a *portion* of an array (actually the pointer) beginning part way through the array, use the ampersand and a specific value in the brackets as in `display(&array[7]);`.

PASSING PARAMETERS

An additional built-in feature of high-level language procedures and functions is the ability to pass parameters and return values (it is ultimately done in assembly code and can be done by the assembly language programmer as well!). As programmers considered the need to avoid repeated code, they realized that much code is *almost* the same. The same thing is done to one variable this time and to a different variable the next time. By using a *separate* variable *in the routine*, filled first with one variable's value and later with another variable's value, the routine acts as though it is always using the same variable. You just fill in the *routine's* variable before calling the routine. C and PL/M do that automatically; it is called **passing parameters**. Both languages have a strictly defined way this is carried out in the assembly code, such as, "the first byte passes in r1, the first word in r1/r2, etc.," but the compiler handles that. (If you mix languages, then see Chapter 6 on modular programming.)

EXAMPLE: SAY NUMBERS

The example in Figs. 5–3 and 5–4 illustrates both passing parameters and use of drivers. It causes a Digitalker (a fixed phrase speech synthesis device of rapidly declining availability) to *speak* the number sent to it. It happens that the code for speaking numbers is sequential from "one" through "twenty" and "thirty" through "hundred" follow. "Zero" is off by itself (Fig. 5–4a shows the codes corresponding to the available words). The example is given in two parts, again illustrating the masking of the details in a *driver*. The digitalker driver is the speak routine. *Nesting* of routines is illustrated because the number conversion routine (Say Numbers in Fig. 5–4) *uses* the digitalker driver routine (Speak in Fig. 5–3). Figure 5–4 also includes a simple main program to endlessly count to 100.

HARDWARE WITH DONE INDICATION: If you interface a device that takes some time to complete an operation (as with an A-D converter or a speech device), test for the done condition at the *start* of the driver rather than the end. Then start the operation and leave the routine. In that way the computer can be doing something else while the hardware finishes and can be right back waiting to start the next operation as soon as the previous one finishes. Only hardware interrupts can be more efficient.

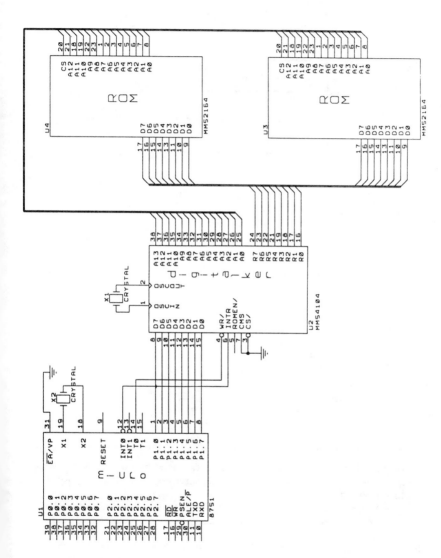

Figure 5–3a Digitalker Schematic

SPEAK

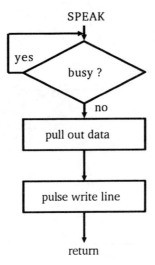

return

Figure 5–3b Speak Routine Flow Chart

```
#include <reg51.h>
#define uchar unsigned char
sbit dgtkbusy=P3^2
sbit dgtkstart=P3^4
                    3
void speak(uchar dgtkcode){
1    while (dgtkbusy);
2    P1=dgtkcode;
     dgtkstart=0; dgtlkstart=1;
}
```

[1]This endless, do-nothing loop waits for the digitalker to be not busy (if it was speaking something before). It is to be used with care because the digitalker may actually show "busy" until it receives its first command.

[2]A *driver* lumps the details in one place. Here the port assignments and handshaking are hidden from the S A Y N routine. If the hardware were changed, only this routine would need changing.

[3]In ANSI standard C, the parameter to be passed can now be defined within the parentheses. In the older manner it would have been an expression after the parentheses before the braces as speak(dgtkcode) uns char dgtkcode; {.

Figure 5–3c Speak Routine—C

```
2   speakr segment code
    rseg speakr
1   speak: jnb P3.2,speak
    mov P1,r2 ;passed code in r2
    clr P3.4
    setb P3.4
    ret
    end
```

[1]This endless, do-nothing loop waits for the digitalker to be not busy (if it was speaking something before). It is to be used with care because the digitalker may actually show "busy" until it receives its first command.

[2]A driver lumps the details in one place. If the hardware were changed, only this routine would need changes.

Figure 5–3d   Speak Routine—ASM51

```
X:DO;
    DECLARE DGTKBUSY BIT AT(0B2H) REGISTER,
    DKTKSTART BIT AT(0B4H) REGISTER;
                    2
    SPEAK:PROCEDURE(DCODE); /*DIGITALKER DRIVER*/
3   DECLARE DCODE BYTE;
1   DO WHILE (DGTKBUSY);
    END;
    P1=DCODE; /*PORT INVERTED*/
    DGTKSTART=0; DGTKSTART=1;
    END SPEAK;
END X;
```

[1]This endless, do-nothing loop waits for the digitalker to be not busy (if it was speaking something before). It is to be used with care because the digitalker may actually show "busy" until it receives its first command.

[2]The parameter passed to the procedure, DCODE, comes after the word PROCEDURE.

[3]In PL/M the parameter passed must be declared *inside* the procedure before the executable instructions. In addition, PL/M 51 requires that *all* passed parameters must be in MAIN (not AUX-ILIARY) memory. That is not to say that the variable transferred to the passing parameter must be so.

Figure 5–3e   Speak Routine—PL/M

Address   Word
  (decimal)

0	THIS IS	36	E	73	CHECK	110	MINUTE
	DIGITALKER	37	F	74	COMMA	111	NEAR
1	ONE	38	G	75	CONTROL	112	NUMBER
2	TWO	39	H	76	DANGER	113	OF
3	THREE	40	I·	77	DEGREE	114	OFF
4	FOUR	41	J	78	DOLLAR	115	ON
5	FIVE	42	K	79	DOWN	116	OUT
6	SIX	43	L	80	EQUAL	117	OVER
7	SEVEN	44	M	81	ERROR	118	PARENTHESIS
8	EIGHT	45	N	82	FEET	119	PERCENT
9	NINE	46	O	83	FLOW	120	PLEASE
10	TEN	47	P	84	FUEL	121	PLUS
11	ELEVEN	48	Q	85	GALLON	122	POINT
12	TWELVE	49	R	86	GO	123	POUND
13	THIRTEEN	50	S	87	GRAM	124	PULSES
14	FOURTEEN	51	T	88	GREAT	125	RATE
15	FIFTEEN	52	U	89	GREATER	126	RE
16	SIXTEEN	53	V	90	HAVE	127	READY
17	SEVENTEEN	54	W	91	HIGH	128	RIGHT
18	EIGHTEEN	55	X	92	HIGHER	129	SS [1]
19	NINETEEN	56	Y	93	HOUR	130	SECOND
20	TWENTY	57	Z	94	IN	131	SET
21	THIRTY	58	AGAIN	95	INCHES	132	SPACE
22	FORTY	59	AMPERE	96	IS	133	SPEED
23	FIFTY	60	AND	97	IT	134	STAR
24	SIXTY	61	AT	98	KILO	135	START
25	SEVENTY	62	CANCEL	99	LEFT	136	STOP
26	EIGHTY	63	CASE	100	LESS	137	THAN
27	NINETY	64	CENT	101	LESSER	138	THE
28	HUNDRED	65	40HZ	102	LIMIT	139	TIME
29	THOUSAND	66	80HZ	103	LOW	140	TRY
30	MILLION	67	20MS QUIET	104	LOWER	141	UP
31	ZERO	68	40MS QUIET	105	MARK	142	VOLT
32	A	69	80MS QUIET	106	METER	143	WEIGHT
33	B	70	160MS QUIET	107	MILE	[2]	
34	C	71	320MS QUIET	108	MILLI		
35	D	72	CENTI	109	MINUS		

[1] Add to words to make plural [2] Do not use codes beyond 143

Figure 5–4a   Digitalker Word List

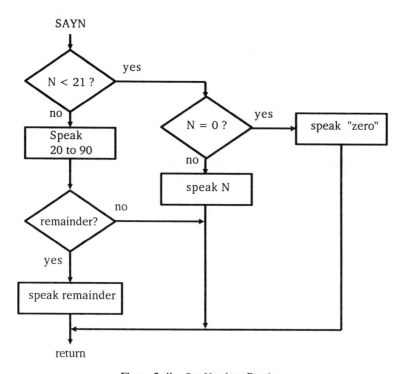

Figure 5–4b Say Numbers Routines

RETURNING VALUES

Besides sending differing parameters *to* a routine, you can get a single value (the "answer" or "result") returned *from* the routine. In PL/M, such a routine is called a **typed procedure**, whereas in C it is a normal **function**. If there are several values to be returned, there is no *direct* way to do it, but it is proper to operate on global variables (discussed next) or to return a *pointer* to an array or structure holding the results. When PL/M or C programs encounter the line, R E T U R N X ;, the routine ends and the single parameter, x, is passed back. Program flow leaves the routine despite any lines that may follow before the E N D or }. With PL/M, the word B Y T E or W O R D indicates that a value is returned. In ANSI standard C, the type of variable to be returned comes before the function name.

V o i d indicates nothing is returned. Void is added to the newer ANSI Standard C so the compiler can check whether you are using the function correctly—earlier versions of C would assume an integer returned and would not check what sort of parameters were being passed. The appearance of r e t u r n x or r e t u r n will terminate the function, but with a void type function, it is more normal to omit the return and just let the function get to the E N D / }.

In C, any routine is called by writing its name (without "call"), and the

```
#define uchar unsigned char
#define uint unsigned
uint i;

void speak(uchar x);

void sayn(uint n){ /*conv 0-99 to speech*/
     uchar x;
     if (n<21){
1      if (n==0)speak(0x1f); /*code for "zero"*/
       else speak(n);   /*"one" through "twenty"*/
     }
     else {       2
       speak((n/10)+0x12); /*"20" thru "90"*/
3      if ((x=n % 10)!=0)speak(x);    /*"1"-"9"*/
     }
}

void main(void){
     for(;;){                     4
       for (i=1;i<=100;sayn(i++));
     }
}
```

[1]The SPEAK routine has the parameter passed as the result of the expression in parentheses.

[2]There is no problem with using an entire expression here since the compiler evaluates it before passing the resultant value to SPEAK.

[3]It would be odd to say "twenty zero." It is possible to avoid carrying out the % operation twice by using an *embedded assignment* to put the result in a temporary variable.

[4]The incrementing of i and calling of sayn can all be embedded in the for loop.

Figure 5–4c   Say Numbers Routines—C

returned result (if any) replaces the name for the evaluation of the rest of the expression where the implicit call (function reference) appeared. Some programmers like to have the words like "call," "begin," and "end" included in their C source to make the code more readable like PASCAL or PL/M. This can easily be done by a set of #define statements where "call" would be defined as nothing, "begin" could convert to {, and "end" could convert to }.

In assembly, returning values is not so automatic, of course, and any value resulting from the use of a subroutine must be left in an agreed upon place such as, *"The 4-byte result will be found in r2/r3,"* or, *"The answer will be found where r0 is pointing."* It is up to the programmer(s) to agree. This is also the case when mixing languages, as will be discussed in Chapter 6.

```
extrn code (speak)

saym segment code
rseg saym
 init: mov r3,#0 ;start of main pgm
 x1: inc r3
    lcall sayn
    cjne r3,#100,x1 ;count to 100
    sjmp init   ;endless loop
sayn:cjne r3,#20,x2 ;pass num in r3
 x2:jnc x3          ;jump if r3>20
    mov a,r3
    jz x4
    mov a,r3
    mov r2,a
    call speak ;say numb (1-19)
    ret
 x4:mov r3,#1fh ;say zero
    ret
 x3:mov a,r3
    mov b,#10
    div ab ;part<10 in b
    add a,#12h
    mov r2,a
    call speak ;say (20-90)
    mov a,b
    jz x5       ;don't say 'twenty-zero'
    mov r2,a
    call speak ;say (1-9)
    x5:ret
end
```

Figure 5–4d Say Numbers Routines—ASM51

EXAMPLE: READ AN A-D CONVERTER

The example of Fig. 5–5 is a routine to drive an 8 or 10-bit A-D converter. The schematics are shown in Fig. 5–5a and Fig. 7–13. The routine starts the conversion, waits until the result is available, reads the result (collecting it byte-by-byte and assembling it into an integer/WORD for the 10-bit version), and then returns the result to the calling function. Multitasking approaches, as discussed in Chapters 8–12, would wait for the result in a way that doesn't tie up the processor.

```
HUNDRED:DO;
  DECLARE I BYTE;

  SPEAK:PROCEDURE (X) EXTERNAL;
    DECLARE X BYTE;
    END SPEAK;

  SAYN:PROCEDURE(N); /*CONV 0-99 TO SPEECH*/
    DECLARE N BYTE,
    IF N<21  THEN DO;
1   IF N=0 THEN CALL SPEAK(01FH); /*"0"*/
    ELSE CALL SPEAK(N); /*"1"-"20"*/
    END;
    ELSE DO;          2
    CALL SPEAK((N/10)+12H); /*"20"-"90"*/
3   IF N MOD 10<>0
      THEN CALL SPEAK(N MOD 10);/*"1"-"9"*/
    END;
    END SAYN;
  DO WHILE 1; /*MAIN PROGRAM*/
  DO I=1 TO 100;
    CALL SAYN(I);
  END;
  END;
  END HUNDRED;
```

[1]The S P E A K routine has the parameter passed as the result of the expression in parentheses.

[2]There is no problem with using an entire expression here since the compiler evaluates it before passing the resultant value to S P E A K.

[3]It would be odd to say "twenty zero." It is possible to avoid carrying out the M O D operation twice if a separate line is added to store the result of the M O D in a temporary variable. The *embedded assignment* of C is not available in PL/M 51 (but is in PL/M 80 as : =).

Figure 5–4e Say Numbers Routines—PL/M

The specific details of using the two different converters are somewhat different because of the particular addressing and the nature of the chips. Notice that all the specific details are hidden in the routine, so changing between devices wouldn't require any changes outside the driver routine. The technique to determine the addresses is discussed further in Chapter 7. You will want to work out these details only *once* for a driver like this one!

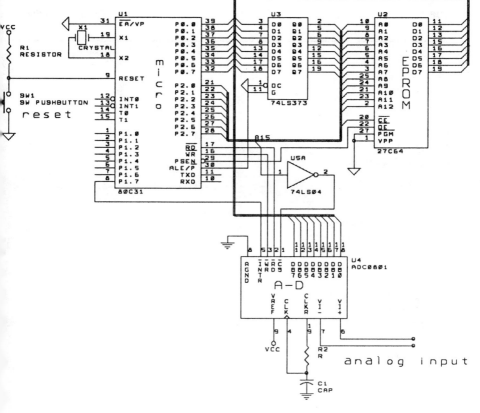

Figure 5–5a A–D Hardware Schematic

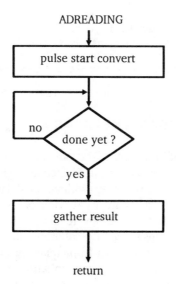

ADREADING

pulse start convert

no — done yet ?

yes

gather result

return

Figure 5–5b A–D Driver Routine

```
1  #include <absacc.h>
   #define uint unsigned int
   #define AD1 XBYTE[0x8000]

   uint ad1rdg (void){
2    sbit ad1busy=P1^7;
     AD1=0; /*cause a start-convert*/
     while (ad1busy); /*wait till done*/
     return AD1;
   }


3  #include <absacc.h>
   #define uint unsigned int
   #define AD2LO XBYTE[0x4000]
   #define AD2HI XBYTE[0x6000]

   uint adreading (void){
     sbit ad2busy=P3^2;
     AD2LO=0; /*cause start-convert*/
     while (ad2busy);
4    return (uint)AD2HI<<8 | AD2LO;
   }
```

[1]This version interfaces the A-D in Fig. 5–5a.

[2]This is a specific method in Franklin C to set the bit at a specific location—here the top bit of P1 is needed, a look at the schematic of Fig. 5–5a.

[3]This version interfaces the A-D in Fig. 7–13.

[4]The *cast* to an unsigned integer (uint) is to ensure that the shifts will not work on an 8-bit value. (It may not anyway depending on the C compiler.) Then the I will be with a 16-bit value so the result will be a 16-bit unsigned int to be returned. This example had considerably more parentheses for safety (which is not a bad idea anyway), but a close check of the precedence rules at the end of Chapter 2 shows that the cast uint precedes the shift << which precedes the or I.

Figure 5–5c A-D Driver Routine—C

EXAMPLE: SCAN A KEYBOARD

The example of Fig. 5–6 may become the heart of a more complete routine. It is a revision of the scan example of Chapter 2 (Fig. 2–7) and uses the same hardware. Here only the scanning (without the delay) is shown. As in Chapter 2, the routine scans the keyboard by driving one column at a time high by putting scanpattern out the top of P1 and reading in the rows from the bottom of the same port. At once it scans successive columns until all four columns have been driven. The result is assembled into a 16-bit quantity, checked for changes (so the calling routine doesn't have those details to worry about), and returned. It *could* have carried out the pushes/releases test described in the Chapter 2 example as well.

```
1   ad1 segment code
    rseg ad1
    adrdg1:mov dptr,#8000h
     movx @dptr,a
    l1: jb P1.7,l1
     movx a,@dptr
     mov r7,a
2    mov r6,#0
     ret
     end


3   ad2 segment code
    rseg ad2
    adrdg2:mov dptr,#4000h
     movx @dptr,a
    l2: jb P3.2,l2    ;end of conversion?
     movx a,@dptr
     mov r7,a
     mov dptr,#6000h
     movx a,@dptr
2    mov r6,a          ;save highest nibble
     ret
     end
```

[1]This works for the A-D of Fig. 5–5a.
[2]The parameter-passing conventions will be discussed in the mixed languages section of Chapter 6. The assumption is that a 2-byte value is returned in r6 and r7.
[3]This works for the A-D of Fig. 7–13.

Figure 5–5d   A-D Driver Routine—ASM51

IN-LINE CODE ALTERNATIVES

Where the time-cost of a routine is unacceptable (perhaps in a very busy system with a short routine that is used frequently), it is possible to use a shorthand notation to represent the instructions in a more understandable way while keeping the benefits of in-line code. In assembly, use **equ**'s (equates), in PL/M **LITERALLY**'s, and in C **#define**'s. In each case the assembler/compiler acts as though the equivalent text is inserted at the time it processes the program, but the *listing* that you see will still have the shorthand name. One additional possibility in assembly is the use of **macro**s. A macro is somewhat like an equate but with the possibility of substituting parameters. The result is in-line code having "blanks" that may be filled in differently each time the macro is invoked. Examples in Chapters 9–11 for DCX and DCX operating systems use a macro in the assembly language for the ITD

```
1  AD1:DO;
     DECLARE AD1 BYTE AT(8000H) AUXILIARY;
     DECLARE AD1BUSY BIT AT(97H) REGISTER;

     AD1READING:PROCEDURE WORD;
     AD1=0; /*STARTS CONVERSION*/
     DO WHILE AD1BUSY;
     END;    /*TEST FOR END OF BUSY */
     RETURN AD1;
     END AD1READING;
   END AD1;

2  AD2:DO;
     DECLARE AD2LO BYTE AT(4000H) AUXILIARY;
     DECLARE AD2HI BYTE AT(6000H) AUXILIARY;
     DECLARE ADBUSY BIT AT(0B2H) REGISTER;

     AD2READING:PROCEDURE WORD;
     AD2LO=0; /*STARTS CONVERSION*/
     DO WHILE ADBUSY;
     END;
3    RETURN SHL(DOUBLE(AD2HI),8) OR AD2LO;
     END AD2READING;
   END AD2;
```

---

[1]This version works with Fig. 5–5a.
[2]This version works with Fig. 7–13.
[3]The DOUBLE forces the value of the port reading to be stored as a WORD, so the shift by 8 will not destroy the reading. Then the OR will be with a 16-bit value so the result will be a 16-bit WORD result to be returned.

Figure 5–5e   A-D Driver Routine—PL/M

(initial task descriptor). There it simplifies the collecting of bytes and nibbles into a predefined format.

SCOPE OF VARIABLES

In a program, the compiler needs to "know about" all the variables in use. Several people may write pieces of the overall program and keep their pieces in separate files. This is called *modular* programming and is discussed in Chapter 6. But what happens if two routines accidentally use the same *name*, say i for an index or x

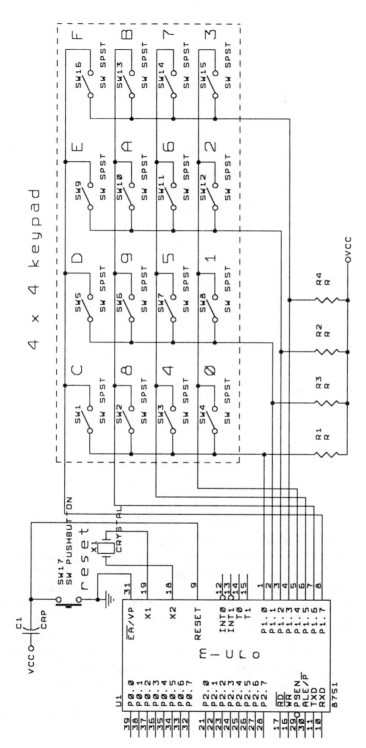

Figure 5–6a Keyscan Hardware

KEYSCAN

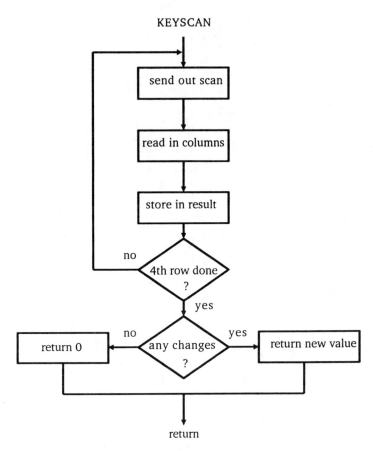

Figure 5–6b   Keyscan Routine

for an unknown, for variables when meaning quite different things stored at different addresses? There can be trouble! The problem is related to the **scope** of the variable. The scope of a variable is the range of program instructions over which its name has meaning. Outside the scope of the variable the name is either undefined or refers to a different variable. There have been definite rules established to avoid confusion. A piece of program outside the *scope* of a given variable cannot refer to it by *name*. Such a routine might accidentally or intentionally do so by using *based*/pointer addressing, but that moves your code out of the safety of the compiler checking. Program revisions might result in the linker putting the variable at a different address and you might forget to change the pointer. Here are the rules within one file (program module):

C recognizes variables found at the top of a file anywhere within the file. Such variables are also known to *other* files, as is discussed in Chapter 6. Tech-

```
#include <reg51.h>
#define uint unsigned int
#define uchar unsigned char
1   uint oldscan;

    uint keyscan(void){
1   uchar pattern;
    uint temp;
    uint newscan=0;
    for (pattern=0x10;pattern>0;pattern<<=1){
      P1=pattern;
      newscan=(newscan<<4) | (P1&0x0f);
    }
2   temp= (newscan != oldscan) ? newscan : 0;
    oldscan=newscan;            3
    return temp;
    }
```

[1]The choice of variable declarations is intentional to allow the variables used only within the procedure to be overlaid. O l d s c a n can be defined outside the function or defined s t a t i c (see Fig. 6–1 for a more detailed discussion of *scope* of variables).

[2]Since o l d s c a n must be updated *before* leaving, and since r e t u r n terminates the routine, it is necessary to use t e m p before r e t u r n.

[3]The ? : operator works like the i f / t h e n / e l s e operations in that the expression in the (unnecessary) parentheses is evaluated and if true, results in the first expression (here n e w s c a n) or if false results in the 0. In this case the resulting expression is assigned to t e m p. Remember that the = works from *right to left* so the assignment to t e m p comes last. Clearly, if you want more directly understood programs, you would avoid this operator or at least add additional parentheses to make the right-to-left more consistent with normal algebra.

Figure 5–6c   Keyscan Routine—C

nically, a variable *outside* a function is known from the point of definition to the end of the file as well as in any other files. The variable need not be defined at the top; in fact, some programming styles now put the main function *first* and define the variables later, but this then requires that the variables be **prototyped** (declared as to their type and number) before they are referenced. **Automatic variables** (defined within a function) are known only to that function. They are subject to overlaying unless the word s t a t i c is included. Parameters passed into a function, of course, must be defined in the procedure and are unknown outside.

Within an assembly language program file, labels (defined with a : colon in the instructions or with an e q u) work *anywhere* in that one file; a first assembler pass is used to pick up all the names. You can't use the same label twice in one file. Tying in with *other* files is the subject of Chapter 6.

In PL/M, any variables declared at the top of the file (just after the required

```
permanentdata segment data
rseg permanentdata
   oldscan: ds 2

keym segment code
rseg keym
keyscan: clr a
     mov r6,a          ;newscan in r6,r7
     mov r7,a
     mov r2,#10h        ;start column 1
ks1: mov P1,r2     ;drive column
     call rot4         ;ready the result
     mov a,P1          ;get result
     anl a,#0fh
     orl a,r7
     mov r7,a
     mov a,r2
     rl a              ;move to next column
     mov r2,a
     cjne r2,#10h,ks1  ;done scanning?
     mov r2,#0         ;use as ='s counter
     mov a,oldscan
     cjne a,6,ks2      ;danger-assumes bank 0
     inc r2            ;one 'no change'
ks2: mov oldscan,r6
     mov a,oldscan+1
     cjne a,7,ks3      ;assumes bank 0
     inc r2            ;one 'no change'
ks3: mov oldscan,r7
     cjne r2,#2,ks4    ;both bytes unchanged?
     clr a
1    mov r7,a
     mov r6,a          ;return 0
ks4: ret

rot4:mov r3,#4     ;rotate r6,r7 left by 4
   ro1:clr c
     mov a,r6
2    rlc a
     mov r6,a
     mov a,r7
     rlc a
     mov r7,a
     jnc ro2
     inc r6                ;carry around
   ro2:djnz r3,ro1   ;4-bit rotate
     ret
end
```

[1]Consistent with PL/M and some versions of C, the two-byte result is returned in r6 and r7.
[2]The rotate left by 4 is put into a subroutine since it is used repeatedly and is a clearly defined operation.

Figure 5–6d   Keyscan Routine—ASM51

```
KEY:DO;
  $INCLUDE(REG51.DCL)
  DECLARE AS LITERALLY 'LITERALLY',
    AUX AS 'AUXILIARY';
  DECLARE OLDSCAN WORD AUX;

  KEYSCAN:PROCEDURE WORD;
1  DECLARE PATTERN BYTE MAIN;
   DECLARE (NEWSCAN,TEMP) WORD MAIN;
   NEWSCAN=0;
   PATTERN=00010000B;
   DO WHILE PATTERN>0;
   P1=PATTERN;
   NEWSCAN=ROL(NEWSCAN,4)OR(P1 AND 0FH);
   PATTERN=SHL(PATTERN,1);
   END;
2  IF NEWSCAN<>OLDSCAN THEN TEMP=NEWSCAN;
   ELSE TEMP=0;
   OLDSCAN=NEWSCAN;
   RETURN TEMP;
   END KEYSCAN;
 END KEY;
```

[1]The choice of variable declarations is intentional to allow the variables used only within the procedure to be overlaid. OLDSCAN must be defined outside the procedure (see Fig. 6-1 for a more detailed discussion of *scope* of variables).

[2]Since OLDSCAN must be updated *before* leaving, and since RETURN terminates the routine, it is necessary to use TEMP before RETURN.

Figure 5–6e   Keyscan Routine—PL/M

DO; that sets up the module) is known throughout that file *except* where there is a procedure that has declared a variable with the *same* name. Within a procedure, any variable declared *in* the procedure is known *only* in that procedure. Thus, even if you declare it outside the procedures, you can use I for an index variable name or CHAR for a variable name without worrying about confusion with the same thing used in another procedure. At a minimum, any parameters passed *to* a procedure must be declared *within* the procedure (in main memory with PL/M 51) and are known only within the procedure.

STYLE SUGGESTIONS

Some programmers put the main function before the other functions, so they have to prototype *all* the functions ahead of the main function. In my opinion, a differ-

DUAL DEFINING A VARIABLE: It is a common error to de-fine/DECLARE a variable as a passed parameter as well as a global (outside the routine) variable. Review the section on the scope of variables. If the variable is global and you want to use it inside a routine, no special definition is needed—just use it by name. The common error is to pass in a variable by the same name as the global variable, make changes to the inside vari-able and expect to have changed the outside variable. Then you wonder why the outside variable (which you think is the same as the inside one) hasn't changed.

FORGETTING OVERLAYING: Any variable declared *within* a procedure is subject to overlaying unless overridden with com-piler directives. Local variables within a procedure cannot be expected to hold the same value when entering the procedure as when the procedure was last exited. For example, an inter-rupt procedure to count the number of timer-driven interrupts to serve as the high-order part of a time count must not declare the count variable within the procedure.

ent approach (*required* in PL/M) is probably best for C: define the variables, fol-lowed by the functions (with the call*ing* ones after the call*ed* ones), and put the main function last. The approach of putting any functions ahead of their use seems easier because it isn't necessary to describe them *twice*—first in the proto-type and later in the definition. The compiler handles this readily, and the main objection must be that the reader encounters the *details* before the *broad overview* of the main program. The only exception is the need to prototype functions found in other modules, as is discussed in Chapter 6.

Be aware that ANSI standard C puts the definition of the passed parameters within the parentheses, whereas earlier versions of C put it after the parentheses but before the body of the function. For compatibility with the earlier versions, the latter can still be done, but this hinders type checking for errors in parameters. If no parameters are passed, the compiler needs to be *told* so by putting the word v o i d within the parentheses. Otherwise it will assume that the old style of C is involved and will omit error checking.

REVIEW AND BEYOND

1. What is the "60-line rule"? What happens when it is violated?
2. What are the proper names for "routine" in the three languages used? Are there any differences between them?
3. What is a "driver," and what are its advantages?
4. Do you use "top-down" programming? Explain your present approach to program development.
5. What can be done when a routine needs to return more than one parameter?
6. When is in-line code preferable to routines?

Chapter 6

Modular
Program Development

This chapter first illustrates modular programming with all the modules in the same language. Then it shows modular programming where the modules are written in different languages and the parameter-passing conventions must be accommodated. A technique to determine these conventions regardless of the compiler in use is shown.

WHY MODULAR?

Modular programming is an approach to software development where individual routines may be separately written, compiled, and debugged. Then the modules are brought together for the final link/locate stage. There are several benefits to modular programming:

1. It makes program development more efficient. Small pieces are easier to understand and test than large programs. When you know the expected inputs to the module and the desired outputs, it is straightforward to test a small module.

2. It is possible to keep routines in a library for future use when the same sort of requirement arises. If you require a display driver again, take it from the library (and modify it if necessary) without starting all over.

3. Modular programming allows the isolation of a problem to a specific module, greatly simplifying debugging.

While illustrating modular programming, this chapter will also outline the software development process, which comes after the initial programming.

PROGRAM PIECES

Several commonly used terms need to be carefully defined before proceeding:

1. A **segment** is a connected unit of code or data memory. Segments may be **relocatable** in the sense that their final locations in memory are left up to the link/locate utility, or **absolute** in the sense that they are specifically assigned by the programmer. For example, the addresses of memory-mapped ports or other hardware addresses are not for the linker to choose. Likewise, a user-configuration table in a *separate* EPROM could not be put just anywhere. A *complete* segment is the combination of the *partial* segments from the various modules. For example, the complete code segment for a multimodule program is the combination of the code segments for the individual routines and the segment for the main program.

2. A **module** is a *file* holding a collection of one or more segments and is given a name by the programmer. Often a module is a collection of related routines for such functions as display, calculation, or user interface.

3. A **library** is a file holding one or more modules. Usually the modules are relocatable object modules from the compiler or assembler to be combined with other modules at link time. The linker selects out of a library *only* the modules referenced by other modules—it puts in only those modules requested by other modules.

4. A **program** is the goal of the entire development process. In this context it is a *single, absolute* module that merges all the absolute and relocatable segments from all the necessary modules. It is ready to run by downloading or burning in EPROM. In other types of computers, programs may be loaded into memory at run time and may be put in *different* locations by the operating system depending on what else is already occupying memory space. This is called "load time locatable."

THE DEVELOPMENT PROCESS

Before going into details of modular programming, it is useful to discuss the program development steps for embedded applications. Figure 6–1 charts these steps for multiple modules, but except for the use of the librarian, the same development process applies. Multiple modules can be combined without the librarian, but the linking instruction line gets messy.

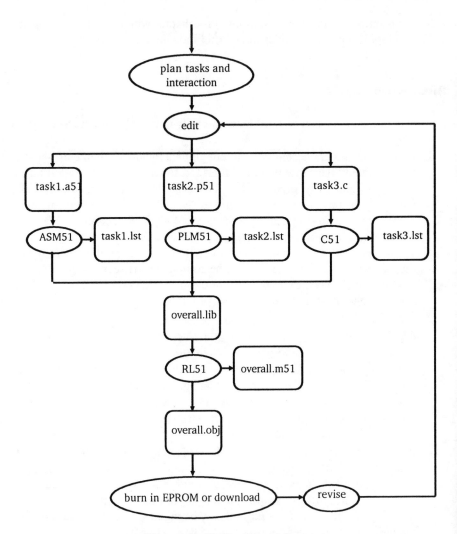

Figure 6–1 The Development Process

The process is often repeated many times, because everything seldom works perfectly at first. Good system developers plan to get pieces working before integrating them into a finished package. Chapter 1 went into more detail on this process of **system integration**, which involves getting the pieces working *together*. Figure 6–1 can be described as follows:

1. **Plan** the overall project, as was emphasized in Chapter 1. Include the particular hardware that will be used, and plan how the software will be divided up to get the job done.

2. **Write** the software, and type it into files for assembling or compiling.

3. **Compile** and/or **assemble** the source programs. This may include putting the object modules into libraries.

4. Get the resulting files (usually **relocatable** so all the jump addresses have not been filled in) set up at specific memory locations. This is **locating** and usually includes **linking**. There are often several pieces to the overall program which have been written separately, or else there are C and PL/M libraries to include.

5. Get the resulting **absolute** file into the computer that will be doing the control job. This is **downloading** if done by a resident monitor program, or **burning** if the file goes into an EPROM for installation on the target computer board.

6. Try out the program to see if there are areas that don't work or need improvement (**debugging**).

FILE NAMING CONVENTIONS

There are several customary extensions for program development files. Typically the source files are `.A51`, `.P51`, and `.C51` or `.C`. The list file, which shows the assembled code and errors, is `.LST`. The (relocatable) object modules are `.OBJ` files, whereas the final (absolute or load time locatable) module is the same name with no extension. If it is converted to Intel's printable format, it is `.HEX`. The libraries are `.LIB` and the "listing" for the linking/locating process is either `.M51` or `.MAP`. Where linking and locating is a two-step process, the linker output is usually `.LNK`.

Additional input files to be included in a source at compile time are `.INC` or, with C, `.H`. Not all C compilers follow these conventions—there is a wide variation, which makes it difficult to generalize.

ASSEMBLER

There are a number of assemblers available for the 8051 family. Many of the C compiler suppliers either include or have available assemblers as well. An assembler takes the mnemonics of the assembly language and translates it into corresponding machine codes. Because there is a one-to-one correspondence, simple assemblers are relatively easy to write. Certain *compilers* produce only assembly mnemonics (assembly language), which must then be assembled to machine code. When it comes to mixing languages, do not automatically assume that one company's assembler will mix with another one's compiler. Franklin's languages seem to mesh well with Intel's languages and tools. In many cases they use the same words and directives. BSO/Tasking's tools are quite different and must be used as an entire package.

EXAMPLE: STEPPER DRIVER—ASM51

Three assembly language program **modules,** s t e p, m s e c, and s t e p r in Fig. 6–2b illustrate a modular program approach to driving a stepper motor much like the one in Chapter 5. The flow charts of Fig. 6–2a describe the three parts of the overall program. The hardware setup would look the same as that of Fig. 5–1a. Figure 6–2c shows the commands that might be used to assemble and then link these modules into an absolute file (here in Intel HEX format). The set of commands invoking the assembler and linker might be put into a batch file and made more general through the use of parameter substitutions—see your DOS manual for more on the use of batch files. Together, the three program pieces will direct-drive a 4-phase stepper motor in full steps at an 8mSec per step rate. The main program (Fig. 6–2b) keeps track of the position of the stepper in phase, puts it into state, and calls stepr in some other module.

In both modules, all the segments but one are **relocatable** by use of the s e g m e n t and r s e g directives. In general, a serious modular programmer will avoid absolute segments and allow the linker to make efficient use of memory spaces. The only absolute address is the XRAM segment that is the I/O port address. Publicly defining absolute address ports in one module with all other modules making them e x t r n can make revisions much easier if the hardware address should be changed some day. The two program pieces can be in *separate* files and still work together because the important labels are made **public** in one and **extrn** in the other. This is discussed more in a later section on the scope of variables. Once you understand the use of names for the segments as well as for the p u b l i c and e x t r n labeling of the variables, relocatable segments are no problem.

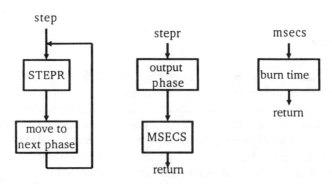

Figure 6–2a   Modular Stepper Programs

```
                                1    extrn code (stepr)
                                2    extrn data (state)
                                3    myvar segment data
                                4    step segment code
                                5    stack segment idata
                                6
      —                         7  1 rseg stack
      0000                       8    ds 10h
                                9
      —                        10    rseg myvar
      0000                      11    phase: ds 1
                               12
      —                        13    rseg step
      0000    750000  F        14    start: mov phase, #0
      0003    758100  F        15  1  mov sp,#stack-1
      0006    850000  F        16  2 loop: mov state, phase
      0009    120000  F        17    lcall stepr
      000C    0500    F        18    inc phase
      000E    E500    F        19    mov a, phase
      0010    5403           20    anl a,#03h
      0012    F500    F        21    mov phase,a
      0014    80F0           22    sjmp loop
                               23    end
```

SYMBOL TABLE LISTING

NAME	TYPE	VALUE		ATTRIBUTES
LOOP C	ADDR	0006H	R	SEG=STEP
MYVAR . . . . D	SEG	0001H		REL=UNIT
PHASE . . . . D	ADDR	0000H	R	SEG=MYVAR
SP . . . . D	ADDR	0081H	A	
STACK I	SEG	0010H		REL=UNIT
START C	ADDR	0000H	R	SEG=STEP
STATE . . . . D	ADDR	—	EXT	
STEP C	SEG	0016H		REL=UNIT
STEPR C	ADDR	—	EXT	

REGISTER BANK(S) USED: 0

[1]The stack is not automatically set to any particular value, and assembly has none of the auto-initialize features of PL/M or C. This example sets the stack to a reserved block of on-chip RAM, which is assigned by the linker.

[2] Note that the use of s t a t e is limited to 0 to 3 in *this* module. It might be better practice to limit it in the subroutine because you might forget and send it something outside the range from another place.

Figure 6–2b   STEP, MSEC, and STEPR Modules

```
LOC OBJ              LINE  SOURCE

                       1   public msec
                       2   msecm segment code
                       3
    ——                 4   rseg msecm
0000 6009              5   msec: jz x ;quit if a = 0
0002 78FA              6      mov r0,#250 ;250 x 4 = 1000
0004 00                7   z: nop
0005 00                8      nop
0006 D8FC              9      djnz r0,z ;4 uSec per loop
0008 D5E0F5           10      djnz acc, msec
000B 22               11   x: ret
                      12   end
```

SYMBOL TABLE LISTING

NAME	TYPE	VALUE	ATTRIBUTES		
ACC .... D	ADDR	00E0H	A		
MSEC C	ADDR	0000H	R	PUB	SEG=MSECM
MSECM C	SEG	000CH			REL=UNIT
X C	ADDR	000BH	R		SEG=MSECM
Z C	ADDR	0004H	R		SEG=MSECM

REGISTER BANK(S) USED: 0

```
LOC OBJ              LINE  SOURCE

                       1   extrn code (msec)
                       2   public state, stepr
                       3   myrom segment code
                       4   myiram segment data
                       5   stppgm segment code
    ——                 6   rseg myrom
0000 0A                7   table: db 00001010b, 00001001b
0001 09
0002 05                8          db 00000101b, 00000110b
0003 06
    ——                 9   reseg myiram
0000                  10   state: ds 1
                      11
    ——                12   xseg at 0ffc0h
FFC0                  13   PORTA: ds 1
                      14
    ——                15   rseg stppgm
0000 900000 F         16   stepr: mov dptr, #table
0003 E500   F         17 3    mov a, state
```

[3] The subroutine module (Fig 6–2b) has a routine, s t e p, which receives a value between 0 and
3. Based on that value (in s t a t e), it outputs one of four table values to PORTA, delays, and returns

Figure 6–2b Continued

```
0005 93          18        movc a, @a+dptr
0006 90FFC0      19        mov dptr, #PORTA
0009 F0          20        movx @dptr, a ;output new phase
000A 7408        21        mov a,#8
000C 120000 F    22 4      lcall msec
000F 22          23        ret
                 24        end
```

SYMBOL TABLE LISTING

NAME		TYPE		VALUE		ATTRIBUTES
MSEC	C	ADDR		—	EXT	
MYIRAM	D	SEG		0001H		REL=UNIT
MYROM	C	SEG		0004H		REL=UNIT
PORTA	X	ADDR		FFC0H	A	
STATE	D	ADDR		0000H	R PUB	SEG=MYIRAM
STEPR	C	ADDR		0000H	R PUB	SEG=STPPGM
STPPGM	C	SEG		0010H		REL=UNIT
TABLE	C	ADDR		0000H	R	SEG=MYROM

REGISTER BANK(S) USED: 0

[4]This routine uses the identical msec routine found in Chapter 3.

Figure 6–2b   Continued

```
1    asm51 step.a51 debug
     asm51 msec.a51 debug
     asm51 stepr.a51 debug
2    rl51 step.obj,stepr.obj,msec.obj
3    oh step
4    copy step.lst+msec.lst+stepr.lst+step.m51
```

[1]These three assembly lines are probably not something to put into a batch file since it takes a good bit of time to reassemble the modules that may not have had any changes or errors. If you were doing this routinely, the use of a library would be highly recommended. The debug causes the symbol and line labels to be included in the object file that results. Without it, the linker/locater will not be able to produce a listing of the final resting places for things.

[2]Again, this might better be done in a library—especially the commonly used routines such as msec. The linker puts the first object file first in the space so the step module ends up at the start of the code space.

[3]This converts the binary object file into Intel's HEX format which is necessary for some PROM programmers and simulators. It is a format that can be sent to a printer directly for manual deciphering of the final code.

[4]It is sometimes useful to combine the list files with the map (.m51) file so they can be viewed or printed together.

Figure 6–2c   Assembling and Linking Commands

```
INPUT MODULES INCLUDED
STEP.OBJ (STEP)
STEPR.OBJ (STEPR)
MSEC.OBJ (MSEC)
```

```
                        LINK MAP FOR STEP (STEP)

   TYPE    BASE    LENGTH      RELOCATION  SEGMENT NAME

   REG     0000H   0008H                   "REG BANK 0"
   DATA    0008H   0001H       UNIT        MYVAR
   DATA    0009H   0001H       UNIT        MYIRAM
   IDATA   000AH   0010H       UNIT        STACK
           0000H   FFC0H                   *** GAP ***
   XDATA   FFC0H   0001H       ABSOLUTE
   CODE    0000H   0016H       UNIT        STEP
   CODE    0016H   0010H       UNIT        STPPGM
   CODE    0026H   000CH       UNIT        MSECM
   CODE    0032H   0004H       UNIT        MYROM


   VALUE           TYPE        NAME

   ——              MODULE      STEP
   C:0006H         SYMBOL      LOOP
   D:0008H         SEGMENT     MYVAR
   D:0008H         SYMBOL      PHASE
   D:0081H         SYMBOL      SP
   I:000AH         SEGMENT     STACK
   C:0000H         SYMBOL      START
   C:0000H         SEGMENT     STEP
   ——              ENDMOD      STEP
   ——              MODULE      STEPR
   D:0009H         SEGMENT     MYIRAM
   C:0032H         SEGMENT     MYROM
   X:FFC0H         SYMBOL      PORTA
   D:0009H         PUBLIC      STATE
   C:0016H         PUBLIC      STEPR
   C:0016H         SEGMENT     STPPGM
   C:0032H         SYMBOL      TABLE
   ——              ENDMOD      STEPR
   ——              MODULE      MSEC
   D:00E0H         SYMBOL      ACC
   C:0026H         PUBLIC      MSEC
   C:0026H         SEGMENT     MSECM
   C:0031H         SYMBOL      X
   C:002AH         SYMBOL      Z
   ——              ENDMOD      MSEC
```

Figure 6–2d   Map File—ASM51

```
:10000000750800758109850809120016050 8E508BC
:060010005403F50880F026
:040032000A090506AC
:10001600900032E5099390FFC0F074081200262282
:0C002600600978FA0000D8FCD5E0F52253
:00000001FF
```

Figure 6–2e   HEX File—ASM51

HEX FILES

The format for the resulting output code files (from Intel and Franklin, at least) is in "Intel HEX format," which is not difficult to decipher. It is a format where all the bytes in the file are printable ASCII characters so they can be viewed. Other, more compact forms, "binary," represent each byte of code with a single byte which allows many of the nonprinting ASCII codes to be in the file.

The colon (:) is the indication of a new record. The next two characters are the Hexadecimal representation for the length of the block in terms of the number of actual data bytes represented. Typically a 10 represents a block of 16 bytes of data. The next four characters are the hex for the 16-bit starting address for the data in the block. The next two characters are a code for the type of block to follow—00 is absolute data and 01 is the end-of-file block. (There are a number of other codes for relocatable files and files related to features of the 8086, for example.) Next comes the actual data where each hex digit pair represents a byte and 16 bytes are represented by 32 characters. The last two digits, easily confused with data, represent the checksum. When all the two-character hex values are added up modulo 256 *with* the checksum, the total should be zero.

COMPILER—PL/M

Compiling for PL/M is not complex. There are several compiler directives that may be put as parameters in the invocation line or may have to be inserted in the source (. P 5 1) file with a $ sign in the first line. Typical are the $ I N C L U D E to insert a file at compile time, and the $ L I S T and $ N O L I S T, which can shut off the insertion of long repeated portions into the listing. Another very useful directive is the $ C O D E / $ N O C O D E directive, which causes the list file to contain the mnemonics of the (assembly) code that results from the compiler. It is particularly useful when one is not sure whether the PL/M is doing the desired thing in some specific part of the program, but it is not something one would usually do for an entire program.

EXAMPLE: STEPPER DRIVER—PL/M

The same sort of example as Fig. 6–2, done in PL/M, is shown in Figs. 6–3a through 6–3d.

```
1   1     STEP:DO;
2   1        DECLARE PHASE BYTE;


3   2 1     STEPR:PROCEDURE(X) EXTERNAL;
4   2           DECLARE X BYTE MAIN;
5   1        END STEPR;


6   1        PHASE=0;
7   2        DO WHILE 1;
8   2           CALL STEPR(PHASE);
9   2           PHASE=(PHASE + 1) AND 03H;
10  2        END;
11  1     END STEP;
```

```
MODULE INFORMATION:      (STATIC+OVERLAYABLE)
   CODE SIZE              = 0012H          18D
   CONSTANT SIZE          = 0000H          0D
   DIRECT VARIABLE SIZE    =  01H+00H      1D+   0D
   INDIRECT VARIABLE SIZE  =  00H+00H      0D+   0D
   BIT SIZE               =  00H+00H      0D+   0D
   BIT-ADDRESSABLE SIZE   =  00H+00H      0D+   0D
   AUXILIARY VARIABLE SIZE = 0000H        0D
   MAXIMUM STACK SIZE     = 0004H          4D
   REGISTER-BANK(S) USED:    0
   13 LINES READ
   0 PROGRAM ERROR(S)
```

```
1   1     STEPM:DO;
2   1        DECLARE PORTA BYTE AT(0CF00H) AUXILIARY;
3   1        DECLARE TABLE (*) BYTE CONSTANT
                (1010B,1001B,0101B,0110B);
```

[1]The passing of PHASE to STATE is done through the parameter-passing mechanism of the procedure, but notice that the STEPR procedure is declared in the sending module in order to describe the parameters to the compiler. The two pieces are compiled *separately* so the compiler has no other way to know how to treat the procedure. The parameter need not have the same name (X is not the same name as STATE), but it must agree as to size, number, and memory type.

Figure 6–3a   STEP and STEPR Modules

```
4 2 1    STEPR:PROCEDURE(STATE) PUBLIC;
5 2         DECLARE STATE BYTE MAIN;
6 2         PORTA=TABLE(STATE);
7 2         CALL TIME(80);
8 1       END STEPR;

9 1           END STEPM;
```

```
MODULE INFORMATION:                    (STATIC+OVERLAYABLE)
   CODE SIZE                    = 0010H         16D
   CONSTANT SIZE                = 0004H          4D
   DIRECT VARIABLE SIZE         =   00H+01H     0D+  1D
   INDIRECT VARIABLE SIZE       =   00H+00H     0D+  0D
   BIT SIZE                     =   00H+00H     0D+  0D
   BIT-ADDRESSABLE SIZE         =   00H+00H     0D+  0D
   AUXILIARY VARIABLE SIZE      = 0000H          0D
   MAXIMUM STACK SIZE           = 0002H          2D
   REGISTER-BANK(S) USED:         0
   12 LINES READ
   0 PROGRAM ERROR(S)
```

Figure 6–3a   Continued

```
1  plm51 step.p51 debug
   plm51 stepr.p51 debug
2  rl51 step.obj,stepr.obj,plm51.lib ol
3  oh step
4  copy step.lst+stepr.lst+step.m51
```

[1]The compiling for each module would probably be done separately and could benefit from the use of a library to simplify the collecting of the modules at the linking stage. Notice that this does not need a module for the time delay TIME because that is already in the standard PL/M library.

[2]The linking is similar to the assembly linkage of the previous example. The only difference is that the time delay comes out of the PL/M library.

[3]This converts the object file to Intel HEX format.

[4]This is a DOS command to combine the list files with the map file to simplify printing.

Figure 6–3b   Compiling and Linking for PL/M

```
INPUT MODULES INCLUDED
  STEP.OBJ(STEP)
  STEPR.OBJ(STEPM)
  PLM51.LIB(?P0040)
  PLM51.LIB(?PIV0R)
```

LINK MAP FOR STEP(STEP)

TYPE	BASE	LENGTH	RELOCATION	SEGMENT NAME
REG	0000H	0008H		"REG BANK 0"
DATA	0008H	0001H	UNIT	?STEP?DT
DATA	0009H	0001H	UNIT	?STEPM?DT?0
IDATA	000AH	0001H	UNIT	?STACK
	0000H	CF00H		*** GAP ***
XDATA	CF00H	0001H	ABSOLUTE	
CODE	0000H	0003H	ABSOLUTE	
CODE	0003H	0012H	INBLOCK	?STEP?PR
CODE	0015H	0010H	INBLOCK	?STEPM?PR
CODE	0025H	0010H	UNIT	?P0040S
CODE	0035H	0009H	UNIT	?PIV0RS
CODE	003EH	0004H	UNIT	?STEPM?CO

SYMBOL TABLE FOR STEP(STEP)

VALUE	TYPE	NAME
——	MODULE	STEP
C:0003H	SYMBOL	STEP
D:0008H	SYMBOL	PHASE
C:0003H	LINE#	1
C:0003H	LINE#	6
C:0006H	LINE#	7
C:0006H	LINE#	8
C:000CH	LINE#	9
C:0013H	LINE#	10
C:0015H	LINE#	11
——	ENDMOD	STEP
——	MODULE	STEPM
C:0015H	PUBLIC	STEPR
C:0025H	SYMBOL	STEPM
X:CF00H	SYMBOL	PORTA
C:003EH	SYMBOL	TABLE
——	PROC	STEPR
D:0009H	SYMBOL	STATE
——	ENDPROC	STEPR
C:0025H	LINE#	1
C:0015H	LINE#	4
C:0015H	LINE#	6
C:001FH	LINE#	7
C:0024H	LINE#	8
C:0025H	LINE#	9
——	ENDMOD	STEPM

Figure 6–3c Map File—PL/M

```
:10000300750800850809120015E508045403F5086E
:0200130080F17A
:04003E000A090506A0
:10001500E50990003E9390CF00F074501200252220
:10002500600D782B02002F007830D8FED5E0F8223D
:03000000020035C6
:0900350075810A75D00002000378
:00000001FF
```

Figure 6–3d   HEX File—PL/M

COMPILER—C

The process for compiling in C is also not complicated for general use, but varies significantly for different C compilers. The use of include files (.H) and the #pragma directive can produce the desired machine-dependent features (that purists hate), but they are very compiler-specific.

EXAMPLE: STEPPER DRIVER—C

The same two example modules as above, done in C, are shown in Figs. 6–4a through 6–4d. If you were sharing *variables,* the variable must be declared extern in one to avoid the definition of two *separate* variables.

```
1          #define uint unsigned int
2          #define uchar unsigned char
3          uchar phase = 0;
4
5  1       stepr (uchar x);
6
7          void main(void) {
8  1        for(;;) {
9  2         stepr(phase=++phase & 0x03);
10 2         }
11 1        }

MODULE INFORMATION:      STATIC    OVERLAYABLE
   CODE SIZE        =      15          —
   CONSTANT SIZE    =       —          —
   XDATA SIZE       =       —          —
   PDATA SIZE       =       —          —
   DATA SIZE        =       1          —
   IDATA SIZE  .    =       —          —
   BIT SIZE         =       —          —
```

Figure 6–4a   STEP, STEPR, and MSEC Modules

```
stmt level      source


  1             #include <absacc.h>
  2             #define uint unsigned int
  3             #define uchar unsigned char
  4             #define PORTA XBYTE[0xffc0]
  5
  6      1      void msec (uint x);
  7
  8             void stepr(uchar state){
  9   1            code uchar table []=
 10   1                       {0x0a,0x09,0x05,0x06};
 11   1            PORTA=table[state];
 12   1  1         msec(8);
 13   1         }
```

```
MODULE INFORMATION:      STATIC    OVERLAYABLE
   CODE SIZE          =    17          —
   CONSTANT SIZE      =     4          —
   XDATA SIZE         =    —           —
   PDATA SIZE         =    —           —
   DATA SIZE          =    —           —
   IDATA SIZE         =    —           —
   BIT SIZE           =    —           —
```

```
stmt level      source

  1             #define uint unsigned int
  2             #define uchar unsigned char
  3      2      void msec(uint x){
  4   1            uchar j;
  5   1            while (x--> 0){
  6   2               for (j=0;j}<125;j++){;}
  7   2            }
  8   1         }
```

```
MODULE INFORMATION:      STATIC    OVERLAYABLE
   CODE SIZE          =    27          —
   CONSTANT SIZE      =    —           —
   XDATA SIZE         =    —           —
   PDATA SIZE         =    —           —
   DATA SIZE          =    —           —
   IDATA SIZE         =    —           —
   BIT SIZE           =    —           —
```

---

[1]The fact that the s t e p r function is not in the first module does not need to be specified, since the compiler knows how to treat it from its *prototype*. Likewise, the *stepr* module can use the m s e c function because it is prototyped in s t e p r.

[2]The msec function is taken directly from the end of Chapter 3.

Figure 6–4a   Continued

```
c51 step.c debug
c51 stepr.c debug
c51 msec.c debug
rl51 step.obj,stepr.obj,msec.obj,c51s.lib ol
oh step
copy step.lst+stepr.lst+msec.lst+step.m51
```

Figure 6–4b   Compiling and Linking for PL/M

```
INPUT MODULES INCLUDED
 STEP.OBJ(STEP)
 STEPR.OBJ(STEPR)
 MSEC.OBJ(MSEC)
 C51S.LIB(?C_STARTUP)
 C51S.LIB(?C_INIT)
```

LINK MAP FOR STEP(STEP)

TYPE	BASE	LENGTH	RELOCATION	SEGMENT NAME
REG	0000H	0008H		"REG BANK 0"
DATA	0008H	0001H	UNIT	?DT?STEP
IDATA	0009H	0001H	UNIT	?STACK
CODE	0000H	0003H	ABSOLUTE	
CODE	0003H	008CH	UNIT	?C_C51STARTUP
CODE	008FH	001BH	UNIT	?PR?_MSEC?MSEC
CODE	00AAH	0011H	UNIT	?PR?_STEPR?STEPR
CODE	00BBH	000FH	UNIT	?PR?MAIN?STEP
CODE	00CAH	0004H	UNIT	?C_INITSEG
CODE	00CEH	0004H	UNIT	?CO?STEPR

SYMBOL TABLE FOR STEP(STEP)

VALUE	TYPE	NAME
——	MODULE	STEP
C:0000H	SYMBOL	_ICE_DUMMY_
D:0008H	PUBLIC	PHASE
C:00BBH	PUBLIC	MAIN
——	PROC	MAIN
——	ENDPROC	MAIN
C:00BBH	LINE#	7
C:00BBH	LINE#	8
C:00BBH	LINE#	9
C:00C7H	LINE#	10
C:00C9H	LINE#	11
——-	ENDMOD	STEP

Figure 6–4c   Map File—C

C:0000H	SYMBOL	_ICE_DUMMY_
C:00AAH	PUBLIC	_STEPR
——	PROC	_STEPR
D:0007H	SYMBOL	STATE
C:00CEH	SYMBOL	TABLE
——	ENDPROC	_STEPR
C:00AAH	LINE#	8
C:00AAH	LINE#	11
C:00B3H	LINE#	12
C:00BAH	LINE#	13
——	ENDMOD	STEPR
——	MODULE	MSEC
C:0000H	SYMBOL	_ICE_DUMMY_
C:008FH	PUBLIC	_MSEC
——	PROC	_MSEC
D:0006H	SYMBOL	X
D:0005H	SYMBOL	J
——	ENDPROC	_MSEC
C:008FH	LINE#	3
C:008FH	LINE#	5
C:009EH	LINE#	6
C:00A9H	LINE#	7
C:00A9H	LINE#	8
——	ENDMOD	MSEC

Figure 6–4c Continued

```
:0300CA000108002A
:0F00BB000508E5085403FFF5081200AA80F22299
:0400CE000A09050610
:1000AA00EF9000CE9390FFC0F07F087E0012008F81
:0100BA002223
:10008F00EF1FAA0670011ED39400EA9400400BE400
:0B009F00FDEDC3947D50E90D80F722B9
:100000000020003787FE4F6D8FD75810802004A02F9
:1000100000BBE493A3F8E493A34003F68001F20845
:10002000DFF48029E493A3F85407240CC8C333C435
:10003000540F4420C8834004F456800146F6DFE4A0
:100040008000B0102040810204080 9000CAE47E0169
:100050009360BCA3FF543F30E509541FFEE493A313
:1000600060010ECF54C025E060A840B8E493A3FA25
:10007000E493A3F8E493A3C8C582C8CAC583CAF0B1
:0F008000A3C8C582C8CAC583CADFE9DEE780BE50
:0100CD000032
:00000001FF
```

Figure 6–4d HEX File—C

LIBRARIES

Object modules can be gathered together in a library for use by the linker at some later time. The librarian utility uses c r e a t e to originate a new library, a d d or r e p l a c e to put an object module in the library, and d e l e t e to remove a module. L i s t can show the modules currently in the library. The librarian will not add a module where another with the same name or another with the same public symbols exists. This avoids any confusion for the linker when it goes looking for unresolved external references. An example of use of the librarian is shown in Fig. 6–8, and Figs. 6–9 and 6–10 go on to show the use of the resulting library.

With C, it is particularly common to use libraries supplied with the compiler, because many math routines are lengthy and only a few of these routines may be needed in any specific application. Unlike interpreted BASIC, good embedded software tools include only what is necessary for the specific application. At linking time, any references to individual math operations will be unsatisfied in the main program modules. The linker will then search through the math libraries for the code to include, say, the floating-point cosine algorithm. There is variation between compiler/linker systems, but any good one will include only the necessary modules. The library modules do not have any particular order, and a reference can be made in one library module to a symbol found in another module.

LINKER/LOCATER

Linking is the process of tying together all the segments in the modules to produce a complete program. It involves identifying all the public symbols (variable, procedure/function, and label names). Once those go into the catalog, the modules that refer to them begin having address locations filled in—the value for the l j m p x , l c a l l x , or m o v d p t r , # x can start to be filled in. *"Start to,"* is the term because a linker does not assign absolute addresses; rather it assigns addresses *relative to* other similar types of code or data. Once all the external RAM data, say, is collected and laid out so no piece of data overlaps another piece, then the locater can figure out the absolute addresses. The way in which segments are fitted together follows a definite sequence. All the partial segments of the same type (code, xdata, etc.) are combined into a single segment. Clearly from 8051 assembly, some segments cannot be put just anywhere due to the nature of the instructions. For example, I N P A G E identifies segments that must fit on one **page** (the same upper 8-bit address) due to the nature of the internal jumps and calls. I N B L O C K segments must fit in a 2048 byte block because of the use of a c a l l. None of the linkers are clever enough to determine if the calls or jumps *within* the module can be handled, even though part of the total module crosses the boundary! The other limitations on relocation are: P A G E—must not cross a 256-byte page boundary, B I T A D D R E S S A B L E—must be put in the internal RAM space that is specifically bit addressable, and U N I T— allows the segment to begin at any byte (or bit for bit variables). PL/M always produces relocatable segments. With Intel's 8051 tools, the linking and locating is one step, so the distinction between the two is not as clear as in other development systems. Although *locating* is done for single-module programs (except absolute assembly modules), *linking* is the heart of modular programming.

The **locater** is a utility that assigns addresses to each segment. Having arranged the partial segments into a full segment in the linking stage, the locate stage fills in the absolute addresses of the segments. Only a few controls are usually needed for a stand-alone program. The symbol tables of Figs. 6–2d, 6–3c, and 6–4c are produced by the linker/locater utility program (with some editing to fit the book better).

OVERLAYING

With the very limited memory space on-chip, the linker/locater will normally reuse locations when they are no longer needed by a routine. That is to say, if one routine neither calls nor is called (even indirectly) from another routine, then one routine will never be running before the other has finished, and the variables can be kept in the exact *same* RAM space, much as you reuse registers. You *could* manually do this sort of thing directly in assembly, but to have the linker manage it is currently possible only in C and PL/M. It can result in considerably less RAM space requirement if there are several unrelated routines. The default in PL/M is NOOVERLAY, so it must be specifically requested. In C you need to check your compiler manual.

SHARING VARIABLES

You need to understand the relationship between variable or routines in separate modules. When compiling one module, the other modules are not available, so the compiler must be given information *about* anything it will use from other modules. Shared variables or routines have the same addresses filled in by the linker/locater utility. This works *only* if the assembler/compiler was able to set up the code to reference the bytes as bytes, integers/words as integers/words, arrays as arrays, and based/pointer as based/pointer. Each language shares differently. Figure 6–5 is a useful summary of the rules that are described in the next two sections.

> FORGETTING OVERLAYS ON AUTOMATIC VARIABLES: A variable declared within a procedure may not be the same the next time the procedure is entered. In some versions of C and PL/M 51 (but not PL/M 80 or PL/M 86) the variables declared within a procedure might have changed because of overlays.

	Assembly	PL/M	C
ariable allocated in THIS routine			
ariable that may be overlaid		Y:PROCEDURE; DECLARE X WORD; END Y;	Y(){ int X;}
ariable that remains between calls		DECLARE X WORD; Y: PROCEDURE; END Y;	Y(){ static int X;}
ariable allocated at outer level THIS module			
riable known in THIS module ONLY	x:ds 2	M:DO; DECLARE X WORD; END M;	static int X;
riable known to other modules	public x x:ds 2	M:DO; DECLARE X WORD PUBLIC; END M;	int X;

riable allocated in some OTHER module	extrn data (x)
	mov dptr, #x
	movx a, @dptr
		DECLARE X WORD EXTERNAL;	extern int X;

utine known in other modules	public y
	y: . . .
		Y: PROCEDURE PUBLIC;
		. . . .
		END Y;
			Y(){
			. . .
			};

utine known in his module ONLY	y: . . .	Y:PROCEDURE;
		. . . .
		END Y;
			static Y(){
			. . .
			};

utine FOUND in ome other module	extrn code (y) lcall y	Y:PROCEDURE EXTERNAL; END Y;	Y();

Figure 6–5 Scope of Variables and Routines

In C, there is a sharp distinction between *declaring* and *defining* a variable or function. A **definition** makes actual storage allocation (analogous to DE-CLARE in PL/M), whereas a **declaration** (prototyping) states the nature of the variable *without* allocating storage. Any local variables defined *within* a function (called **automatic** variables) may be overlaid and changed between function calls. Some compilers keep them on the stack so they are definitely gone while others (for the 8051) observe the nesting sequence of function calls and put them at the same fixed places for several functions. The only exception for variables defined within a function is for **static** definitions, which keep the value both private and unchanged between function calls. Automatic variables have meaning *only* in the specific function. Variables defined or declared within one file (module) *outside* any function and *before* the functions that use them can be shared among the functions. Variables defined in a module, outside any function but *after* a function that would use them, and variables defined in another module, must have the word **extern** to cause a declaration rather than a definition. Essentially, e x t e r n notifies the compiler to look elsewhere for the variable rather than to allocate storage. So e x t e r n stops the allocation of space. Variables defined *outside* any function are publicly available to *any* function in *any* module. **Static** *outside* any function, however, sets up a shared variable—shared for functions within that module but unrecognized in any other module. This is a convenient way to share a variable within a module, which involves several related functions, without the chance that some unrelated module may stumble on the same name for a variable. The intermodule, global nature of C can be a problem if used carelessly.

In assembly, any variable, subroutine, or other label to be shared with other modules must be identified **public** in the module where it is defined (at the top of the module). Any module that *uses* it must have it included in the e x t r n lines at the top of the module.

In PL/M, each module must have a name and include an outer DO/END block as discussed in Chapter 5. Any variables at the top of this module are **global** to this module. Any variables used *only* within this module can be declared in the usual manner. Any variables declared in this module that may be shared with *other* modules must have the word **PUBLIC** after the declaration. That causes the name to be put in the cross reference table used by the linker. Assuming that the variable is not declared with the AT ( ), the actual address will be determined at locate time, and the variable will be assigned space with other variables declared in this module. If this module *uses* a variable declared in *another* module, this module must declare it, but put the word **EXTERNAL** after it. In this way the compiler can set up the right code to access it, but the location will be filled in later. Within one module variables declared at the outer level will be shared without the PUBLIC/EXTERNAL declaration because their scope is the entire module. Variables declared *within* a procedure are private to that procedure and cannot be used anywhere outside. Those variables have the scope of just that procedure, so it is possible for several procedures to all use, say, I for the index. If a variable is to remain unchanged, it should be declared globally at the top of the module to avoid being overlaid. It is uncommon to declare variables public inside a procedure.

SHARING SUBROUTINES/PROCEDURES/FUNCTIONS

Having considered variables, it is a short step to procedures or functions. Refer again to Fig. 6–5 for a summary of the scope of various objects.

Functions in C are naturally global or public and can come after calling routines. If a function is to be private to one module, then it can be defined **static** and will not be callable from any other module.The ANSI standard for C *suggests* that *all* functions be **prototyped** ahead for the *main* function. Then the actual functions come *after* the main function or in another module. This is just the reverse of PL/M and makes for more program lines, but it does fit the idea of top-down programming discussed in Chapter 1. However, putting the actual function *definitions before* the main function always works and avoids some unnecessary program lines.

In assembly, subroutines are available anywhere within a given module by the label. An assembler makes a first pass to gather up all the symbol names so the values can later be filled in for the l c a l l or l j m p. A subroutine is shared with *other modules* by making the name p u b l i c and e x t r n as shown in Figs. 6–2a,b. When specifying e x t r n, the type of symbol (c o d e , d a t a , x d a t a , i d a t a , b i t , or n u m b e r) must be specified so the linker can be sure the right types of things are together.

In PL/M, within a module any procedure will be known (callable) to any part of the program that *follows* it. Thus, a procedure called from another procedure *must* come *before* the calling procedure. If a procedure is to be recognized in other modules, then it must be made public by putting the word P U B L I C after the P R O C E D U R E line. Then any other module must have the same P R O C E D U R E declaration (without the internal details except the parameter declarations) and the word E X T E R N A L. Within a single module there is no way to put a procedure after its call, which, of necessity, puts the main program last.

MIXING LANGUAGES

When combining program parts written in *different* languages, it is most common to write the hardware-related routines in assembly. There the details are important, and compilers have not historically made efficient use of internal registers. A compiler might tuck away intermediate results in off-chip RAM, simply to pull them out a few lines later because it doesn't look that far ahead. It is also more common to write the main routines in a high-level language where ease of understanding by the reader is more important, and the code is easier to write. A few bytes of extra code used only once cost very little in run-time efficiency, but a few bytes used repeatedly in a loop can be significant. A good approach is to write everything in a high-level language with the c o d e option turned on for the frequently used procedures. When all is working, go back and "fine tune" the critical routines. This is particularly easy with some C compilers that produce only assembly language and in turn must be assembled to get the final code. Some compilers also allow in-line assembly code to lump those routines in the same file. One could even toggle back and forth between languages depending on the whim of the moment; *that* would shake off the casual reader!

PARAMETER-PASSING CONVENTIONS

In mixing languages, the key feature is the passing of parameters and returning values. There has to be complete agreement or else the passed parameter won't be picked up in the routine. The same conventions must be used in both languages. In assembly, of course, the programmer has complete control, so usually the assembly module yields to the high-level language. Unfortunately, each C compiler uses *different* conventions and those even change depending on the choice of the large, medium, or small memory model. Not all compilers can mix size models even among their *own* modules.

So what are these conventions, typically? For PL/M (and optionally for Franklin C) all parameters are passed *to* a routine at a *fixed* set of locations in internal RAM. If bits are passed, they too must be a sequential string of bits located in internal bit-definable space. Of course, the order and size (byte/word: char/int) must be consistent between the call*ing* and the call*ed* routine. Essentially an identically labeled block of internal RAM is shared between the two. The caller fills in the block with the parameters to be passed before issuing the assembly call. The called routine goes ahead with the assumption that the desired values are already in the block when the call comes.

The returning convention always involves only a single variable. It is again the same for PL/M (as far as it goes) and Franklin C, but different for some other C compilers. Figure 6–6 shows this convention for Franklin C and PL/M.

For most general-purpose C compilers as well as *some* 8051-compilers, the prevalent passing convention is by a stack. That is more consistent with C in

RETURN VALUE	REGISTER	MEANING
bit	C flag	
byte/char	r7	
word/int	r6,r7	msb in r6,lsb in r7
long	r4-r7	msb in r4,lsb in r7
float	r4-r7	32 bit IEEE format
pointer	r1-r3	msb r2,lsb r1,selector r3

Figure 6–6   Routine Return Conventions

general and preserves reentrancy. If a function calls itself, the stack just gets deeper rather than writing over its variables. Although such an approach is more general, for the 8051 it is much less efficient because, to access external memory where a large stack could be kept, all activity must be done with one pair of instructions, which requires setting up and saving the data pointer each time. In the internal mode, compilers can use the normal *internal* stack, but that becomes impractical with math library functions, which may consume 100 bytes of stack out of the 128 or 256 available! Other software may need the on-chip space. With DCX, for example (discussed in Chapters 8–12), the consumption of such space would leave no room for messages.

The Franklin C compiler (version 3) can pass parameters using registers, using fixed memory locations like PL/M, or using a stack. Some other compilers have at least some of those options and a few are discussed in Appendix C. There is *no* portability between C compilers for the 8051 family at the *code* level and there are a few nonstandard additions to ANSI standard C in each compiler that have to be changed when changing compilers. Only the C *source* code can be switched across compilers, and even then some of the non-ANSI-standard additions will have to be changed. Any assembly module written to line up with one compiler probably will not match another (note the two versions of Fig. 6–7!). Only a few of the C compilers can interface with PL/M without some assembly language buffer in between.

The module pairs in the first examples (Figs. 6–2 through 6–4) were designed to be compatible within the same language. Franklin's C compiler allows the inclusion of the word **alien** when prototyping a function to signify that the PL/M conventions are to be used in calling it. Obviously assembly can be written to adhere to any desired convention. In the worst case it might be necessary, when mixing languages, to write an assembly language *interface* between the other languages. Several of the older, stack-based C compilers may be quite messy to combine with other languages.

COMPATIBILITY BY TEST

In the end, the easiest way to find out how a given C compiler passes parameters is to compile a dummy function and function call with the c o d e list option turned on. Then you can see exactly what assembly code is produced and can model it in your own call or routine. Figures 6–7a through 6–7g illustrate such a test for PL/M and numerous variations of C parameter passing.

Where the passing conventions of the earlier examples are compatible, the linking is as simple as picking the object module names. Realize that even with compatible parameter-passing conventions, not all companies' software tools produce compatible object file formats; you may be limited to the same company's compiler/assembler when mixing with their other languages. In general, you are advised to obtain all the tools from the same source or have good guarantees of compatibility before you spend your money.

```
1   TST:DO;           1
2   PROC:PROCEDURE(BYT,WD) WORD;
3     DECLARE BYT BYTE, WD WORD;
4     RETURN 0CCCCH;
5   END PROC;
6   DECLARE X WORD;
7   X=PROC(0AAH,0BBBBH);
8   END TST;


2     ; PROCEDURE TST (START)
      ; PROCEDURE PROC (START)
      0000  7ECC      MOV    R6,#0CCH
      0002  7FCC      MOV    R7,#0CCH
      0004  22        RET
      ; PROCEDURE PROC (END)
      0005  7500AA F MOV     BYT,#0AAH
      0008  7500BB F MOV     WD,#0BBH
      000B  7500BB F MOV     WD|0001H,#0BBH
      000E  1100   F ACALL PROC
      0010  8E00   F MOV     X,R6
      0012  8F00   F MOV     X+0001H,R7
      ; PROCEDURE TST (END)
```

[1]This is the source code with statement numbers added. Actually several things that result from the compilation and linking have been stripped out to save space here. One thing omitted is the level of nesting which is another column in the list file.

[2]This is the result with the code option turned on. It comes from the *compiling* so the actual addresses of some things have yet to be determined by the linker. That is designated by the F in some rows. Some lines of PL/M do not directly produce any code. The passing of parameters *to* the procedure will always be in fixed locations in on-chip RAM with PL/M 51.

Figure 6–7a   Parameter Passing Test—PL/M

EXAMPLE: ASSEMBLY LINKED TO PL/M OR C

An example of an assembly module interfaced to another language, a 4-byte math library written by the author, is shown in Fig. 6–8. It is a set of specific routines originally designed to link with PL/M for a predefined calculation where 16 bits was not sufficient. The first part, Fig. 6–8a, has two versions due to a differing parameter-passing convention in the PL/M and the C examples. The difference relates to naming and to passing in of parameters. Incidentally, there would be an additional difference for other C compilers that don't return using r6,r7. A usage example with PL/M is shown in Fig. 6–9, with the same example in C shown in Fig. 6–10.

```
1                              #define uint unsigned int
2               #define uchar unsigned char
3       1       alien uint fn(uchar chr,uint intg){
4    1            return 0xcccc;
5    1          }
6               uint x;
7               main(){
8    1            x=fn(0xaa,0xbbbb);
9    1                    }


                 ; FUNCTION fn (BEGIN)
0000 7ECC            MOV       R6,#0CCH
0002 7FCC            MOV       R7,#0CCH
0004                 ?C0001:
0004 22              RET
                 ; FUNCTION fn (END)
                 ;FUNCTION main (BEGIN)
0000 7500AA    R     MOV       ?fn?BYTE,#0AAH
0003 7500BB    R     MOV       ?fn?BYTE+01H,#0BBH
0006 7500BB    R     MOV       ?fn?BYTE+02H,#0BBH
0009 120000    R     LCALL     fn
000C 8E00      R     MOV       x,R6
000E 8F00      R     MOV       x+01H,R7
0010 22              RET
                 ; FUNCTION main (END)
```

¹This is a special case where the C compiler is told to use the *alien* parameter conventions which are the same ones *always* used by PL/M. It is available for compatibility and is used in the last example in this chapter where both PL/M and C modules are linked to the same assembly module. Since PL/M can't be changed, the C is adaptable! In general this would not be the preferred method of passing for C.

Figure 6–7b   " Alien" Parameter Passing Test—C

CODE EFFICIENCY

Since mixed languages are generally used to improve program efficiency, it is time to reconsider the issue. You need to decide what you mean by efficient code. Is it the code that takes up the least memory, the code that runs in the least instruction cycles, or the code that takes the least time and effort to write? Fortunately, those objectives are not mutually exclusive! In general, code that takes fewer bytes will probably run faster, but the choice of looping versus straight line code can make a difference. It is often said that assembly coding can be more efficient than high-level programming, but that is referring to the size and speed of the final code rather than the time it takes to write and debug it!

All the talk of efficiency assumes you have barely enough processor time to

```
1   #define uint unsigned int
2   #define uchar unsigned char    1
3   uint fn(uchar chr,uint intg){
4     return 0xcccc;
5   }
6   uint x;
7   main(){
8     x=fn(0xaa,0xbbbb);
9   }
```

```
2     ;FUNCTION          _fn (BEGIN)
      0000 8F00      R   MOV      chr,R7
      0002 8C00      R   MOV      intg,R4
      0004 8D00      R   MOV      intg+01H,R5
      0006 7ECC          MOV      R6,#0CCH
      0008 7FCC          MOV      R7,#0CCH
      000A               ?C0001:
      000A 22            RET
        ; FUNCTION        _fn (END)
        ; FUNCTION        main (BEGIN)
      0000 7FAA          MOV      R7,#0AAH
      0002 7DBB          MOV      R5,#0BBH
      0004 7CBB          MOV      R4,#0BBH
      0006 120000    R   LCALL    _fn
      0009 8E00      R   MOV      x,R6
      000B 8F00      R   MOV      x+01H,R7
      000D 22            RET
        ; FUNCTION main (END)
```

[1]This is the source code with statement numbers added. Actually several things that result from the compilation and linking have been stripped out to save space here. One thing omitted is the level of nesting, which is another column in the list file.

[2]This is the result with the code option turned on. It comes from the *compiling* so the actual addresses of some things have yet to be determined by the linker. That is designated by the F in some rows. Some lines of PL/M do not directly produce any code. The passing of parameters *to* the procedure will always be in fixed locations in on-chip RAM with PL/M 51.

Figure 6–7c   Register-based Parameter Passing Test—C

get everything done, or that it matters how quickly the processor finishes some task. In reality, most embedded systems spend the majority of their processor time waiting for something to do. This will be shown in greater detail in Chapter 8 where multitasking is discussed, but it is obvious that a program that toggles a bit every second will, if it runs faster or more efficiently, just spend more time waiting for the next toggle time. You really should consider your specific application before deciding how much effort to put into efficiency. In general, you

```
 1   #pragma NOREGPARMS     1
 2   #define uint unsigned int
 3   #define uchar unsigned char
 4   uint fn(uchar chr,uint intg){
 5    return 0xcccc;
 6   }
 7   uint x;
 8   main(){
 9    x=fn(0xaa,0xbbbb);
10   }


  ; FUNCTION fn (BEGIN)
  0000 7ECC        MOV      R6,#0CCH
  0002 7FCC        MOV      R7,#0CCH
  0004             ?C0001:
  0004 22          RET
  ; FUNCTION fn (END)
  ; FUNCTION main (BEGIN)
  0000 7500AA R    MOV      ?fn?BYTE,#0AAH
  0003 7500BB R    MOV      ?fn?BYTE+01H,#0BBH
  0006 7500BB R    MOV      ?fn?BYTE+02H,#0BBH
  0009 120000 R    LCALL    fn
  000C 8E00   R    MOV      x,R6
  000E 8F00   R    MOV      x+01H,R7
  0010 22          RET
  ; FUNCTION main (END)
```

[1]This is the same as Figure 6–7b *except* the **NOREGPARMS** is turned on. Notice that the main function loads the parameters into on-chip RAM at fixed (to be determined by the link/locate utility) locations rather than into the registers in the previous example.

Figure 6–7d   Fixed On-chip Parameter Passing Test—C

ought to write in high-level language for things that are not repeated often and involve user interaction. If you wish, you can go to assembly for small, tight loops that are repeated often. It is there that improvements in efficiency are multiplied by the repetition of the loop. Probably those loops involve either hardware or specific math operations, which can perhaps be done more specifically than in the libraries (as in Fig. 6–8).

Now that you have seen some examples of compiling with the code option turned on, you can carry out your own investigations of efficiency. There are such wide variations between resultant code depending on the choice of compiler that it is probably useless to get specific. If you study the code, you will see that high-level languages follow a fairly formal set of rules about retrieving and returning variables between lines. Depending on the compiler, there will be more or less

```
1 #pragma NOREGPARMS
2       #define uint unsigned int
3       #define uchar unsigned char
4       uint fn(uchar chr, uint intg)large{  1
5  1    return 0xcccc;
6  1  }
7       uint x;
8       main(){
9  1    x=fn(0xaa,0xbbbb);
10  1  }
```

```
                    ; FUNCTION fn (BEGIN)
0000 7ECC               MOV     R6,#0CCH
0002 7FCC               MOV     R7,#0CCH
0004                ?C0001:
0004 22                 RET
                    ; FUNCTION fn (END)


                    ; FUNCTION main (BEGIN)
0000 900000    R        MOV     DPTR,#?fn?BYTE
0003 74AA               MOV     A,#0AAH
0005 F0                 MOVX    @DPTR,A
0006 A3                 INC     DPTR
0007 74BB               MOV     A,#0BBH
0009 F0                 MOVX    @DPTR,A
000A A3                 INC     DPTR
000B F0                 MOVX    @DPTR,A
000C 120000    R        LCALL   fn
000F 8E00      R        MOV     x,R6
0011 8F00      R        MOV     x+01H,R7
0013 22                 RET
          ; FUNCTION main (END)
```

[1]With the Franklin C compiler, at least, it is possible to force the parameters of fn to be in off-chip RAM by either using the large memory model for the specific function as shown here, or using the large model overall. Then the parameter loading before the function call would be more complex.

Figure 6–7e   Off-chip Parameter Passing Test—C

foresight in evidence relative to keeping things in registers or on-chip RAM between program lines. A large amount of the code may be moving things around. If you write in assembly, you can carefully plan out the register usage to maximize efficiency because you can look far enough ahead to see that some things will be needed again. The compiler may not do that.

Probably it is with efficiency that the choice of C compiler has the biggest

```
1        #define uint unsigned int
2             #define uchar unsigned char
3             uint fn(uchar chr,uint intg)reentrant{   1
4   1             return 0xcccc;
5   1         }
6             uint x;
7             main(){
8   1             x=fn(0xaa,0xbbbb);
9   1         }


             ; FUNCTION _?fn (BEGIN)
0000 1500    E      DEC     ?C_IBP
0002 1500    E      DEC     ?C_IBP
0004 A800    E      MOV     R0,?C_IBP
0006 A604           MOV     @R0,AR4
0008 08             INC     R0
0009 A605           MOV     @R0,AR5
000B 1500    E      DEC     ?C_IBP
000D A800    E      MOV     R0,?C_IBP
000F A607           MOV     @R0,AR7
0011 7ECC           MOV     R6,#0CCH
0013 7FCC           MOV     R7,#0CCH
0015                ?C0001:
0015 0500    E      INC     ?C_IBP
0017 0500    E      INC     ?C_IBP
0019 0500    E      INC     ?C_IBP
001B 22             RET
             ; FUNCTION _?fn (END)
             ; FUNCTION main (BEGIN)
0000 7CBB           MOV     R4,#0BBH
0002 7DBB           MOV     R5,#0BBH
0004 7FAA           MOV     R7,#0AAH
0006 120000  R      LCALL   _?fn
0009 8E00    R      MOV     x,R6
000B 8F00    R      MOV     x+01H,R7
000D 22             RET
             ; FUNCTION main (END)
```

[1]The function can be made reentrant to force parameter passing by a stack. With the small memory model (in use here by default), the stack will be in an on-chip stack. Notice that it is not the actual hardware-supported stack using push and pop.

Figure 6–7f   Reentrant On-chip Parameter Passing Test—C

```
1      #define uint unsigned int
2                  #define uchar unsigned char
3                  uint fn(uchar chr,uint intg)large reentrant{ 1
4      1           return 0xcccc;
5      1           }
6                  uint x;
7                  main(){
8      1            x=fn(0xaa,0xbbbb);
9      1           }


                   ; FUNCTION _?fn (BEGIN)
0000 90FFFE            MOV       DPTR,#0FFFEH
0003 120000   E        LCALL     ?C_ADDXBP
0006 EC                MOV       A,R4
0007 F0                MOVX      @DPTR,A
0008 A3                INC       DPTR
0009 ED                MOV       A,R5
000A F0                MOVX      @DPTR,A
000B 90FFFF            MOV       DPTR,#0FFFFH
000E 120000   E        LCALL     ?C_ADDXBP
0011 EF                MOV       A,R7
0012 F0                MOVX      @DPTR,A
0013 7ECC              MOV       R6,#0CCH
0015 7FCC              MOV       R7,#0CCH
0017               ?C0001:
0017 900003            MOV       DPTR,#03H
001A 120000   E        LCALL     ?C_ADDXBP
001D 22                RET
                   ; FUNCTION _?fn (END)
                   ; FUNCTION main (BEGIN)
0000 7CBB              MOV       R4,#0BBH
0002 7DBB              MOV       R5,#0BBH
0004 7FAA              MOV       R7,#0AAH
0006 120000   R        LCALL     _?fn
0009 8E00     R        MOV       x,R6
000B 8F00     R        MOV       x+01H,R7
000D 22                RET
                   ; FUNCTION main (END)
```

[1]Here is an example that uses an off-chip artificial stack for parameter passing. Notice that it calls in an external subroutine to actually manipulate the stack. Also, it is not clear where this stack will be located. Why is there the loading of dptr with #0FFFE? I would determine where the stack ends up (by disassembling the final located code) before trusting the routine completely.

Figure 6–7g   Reentrant Off-chip Parameter Passing Test—C

1

```
    public dadd,dsub,dmul,ddiv,dload,_dload
    public dresult,?dload?byte    2      4
    dloddat segment data
    dstacka segment data
    dstackb segment data
    dstackc segment data
    dimath segment code

2   rseg dloddat
    ?dload?byte:
    Ain: ds 2 3 rseg dstacka
    STKa:ds 4
    rseg dstackb
    STKb:ds 4
    rseg dstackc
    STKc:ds 4

    rseg dimath
    dload:mov r0,#STKb
      mov r1,#STKc
      acall qmov ;B to C
      mov r0,#STKa
      mov r1,#STKb
      acall qmov ;A to B
      mov r0,#STKa
      clr a ;zero upper 2
      mov @r0,a
      inc r0
      mov @r0,a
      inc r0
```

[1]The key interface routines are d l o d (which picks up the two bytes of A i n and pushes them on to an arbitrary internal stack of three 4-byte values) and d r e s u l t (which pops the arbitrary stack and leaves off the lower two bytes in r6 and r7 t o satisfy the return convention of PL/M). The routines were developed before C was available to satisfy a specific need for a double precision algorithm to extrapolate and scale a set of data points. It would be tempting now to use the C library instead and compare efficiency!

[2]PL/M51 always takes in parameters as fixed locations in MAIN (on-chip) memory. Therefore, the assembly routine sets up a fixed location variable (here called A i n). The naming convention shown for the segment is also necessary to satisfy the PL/M portion of the code—that's what the PL/M compiler is going to name it so the assembly had better provide it!

[3]The arbitrary 4-byte stacks for this math routine could be put into one segment quite nicely, but an early application was using scattered fragments of main memory and these smaller segments could be moved around by the linker to "pack" better.

[4]In the category of name choices, the Franklin C compiler insists on having the underline as the first character of any function that passes parameters by register. Only with #pragma NOREGPARAMS or extern alien will it go to the fixed location passing and drop the underline from the name. Incidentally, for reentrant functions, the naming convention is _?function for Franklin. To add to the fun, some other compilers put the underline in front of everything. Needless to say, some trials are recommended.

```
5  mov @r0,Ain ;load A
      inc r0
      mov @r0,Ain+1
      ret
4
   _dload:mov r0,#STKb
      mov r1,#STKc
      acall qmov ;B to C
      mov r0,#STKa
      mov r1,#STKb
      acall qmov ;A to B
      mov r0,#STKa
      clr a ;zero upper 2
      mov @r0,a
      inc r0
      mov @r0,a
      inc r0
5     mov @r0,6        ;comes in r6,r7
      inc r0
      mov @r0,7
      ret

   ;puts one 4-byte stack value
   ;over another one—moves it
   qmov:mov r2,#4
   qmo1:mov a,@r0     ;from @r0
      mov @r1,a       ;to @r1
      inc r0
      inc r1
      djnz r2,qmo1
      ret

   dresult:mov r7,STKa+3 ;pops stack
      mov r6,STKa+2    ;result in r6,r7
      mov r0,#STKb
      mov r1,#STKa
      acall qmov ;B over A
      mov r0,#STKc
      mov r1,#STKb
      acall qmov ;C over B
      ret
```

[5]The actual difference in the parameter passing shows up here by the use of Ain or r6,r7.

Figure 6–8a   Continued

```
dadd:mov r0,#STKa+3 ;lsb of A
    mov r1,#STKb+3 ;lsb of B
    mov r2,#4       ;4 bytes to process
    clr c           ;no carry into first add
dad1:mov a,@r0
    addc a,@r1      ;A+B
    mov @r0,a       ;save in A
    dec r0          ;move to next higher byte
    dec r1
    djnz r2,dad1    ;4 times
    mov r0,#STKc
    mov r1,#STKb
    acall qmov      ;move C over B
    ret

dsub:mov r0,#STKa+3 ;lsb of A
mov r1,#STKb+3      ;lsb of B
mov r2,#4           ;4 bytes to process
clr c               ;no borrow from first subtract
dsu1:mov a,@r0
    subb a,@r1      ;A-B
    mov @r0,a       ;save in A
    dec r0          ;move to next higher byte
    dec r1
    djnz r2,dsu1    ;4 times
    mov r0,#STKc
    mov r1,#STKb
    acall qmov      ;move C over B
    ret

dmul: mov a,STKa+3
    mov b,STKb+3
    mul ab          ;Al x Bl
    mov r2,a        ;temp byte 0
    mov r3,b        ;temp byte 1
    mov a,STKa+2
    mov b,STKb+2
    mul ab          ;Ah x Bh
    mov r4,a        ;temp byte 2
    mov r5,b        ;temp byte 3
    mov a,STKa+2
    mov b,STKb+3
    mul ab          ;Ah x Bl
    acall addin
    mov a,STKa+3
    mov b,STKb+2
    mul ab          ;Al x Bh
```

Figure 6–8a continued

```
        acall addin
        mov STKa+3,r2    ;temp results into A
        mov STKa+2,r3
        mov STKa+1,r4
        mov STKa,r5
        mov r0,#STKc
        mov r1,#STKb
        acall qmov        ;put C over B
        ret


addin:add a,r3           ;low part from mul ab
      mov r3,a
      mov a,b
      addc a,r4          ;high part from mul
      mov r4,a
      clr a
      addc a,r5          ;if any carry up
      mov r5,a
      ret
end
      ;if no borrow then puts 1 in bottom of A
      ;then shifts A left by 1
      ;answer fills in bottom of A as it shifts
   ddiv:mov r2,#17       ;A(32 bit)/B(lower 16 bits)
   ddi1:clr c
      mov a,STKa+1
      subb a,STKb+3
      mov r3,a
      mov a,STKa
      subb a,STKb+2
      jc ddi2            ;too small—don't use
      mov STKa,a
      mov STKa+1,r3
   ddi2:cpl c
      mov r0,#STKa+3
      mov r3,#4
   ddi3:mov a,@r0        ;shift A left one bit
      rlc a
      mov @r0,a
      dec r0
      djnz r3,ddi3
      djnz r2,ddi1
      mov r0,#STKc
      mov r1,#STKb
      acall qmov
      ret
```

Figure 6–8a Continued

```
      asm51 dimath.a51 debug
1     lib51 create dimath.lib
      lib51 delete dimath.lib(dimath)
      lib51 add dimath.obj to dimath.lib
```

[1]The three lines here show the use of the librarian in a batch file. From the keyboard, the use is a bit simpler in that the l i b51 doesn't come with each command; once you enter l i b51 the librarian prompts with a * and you type just the rest of the line. When done, you leave by typing e x i t. The 3 lines above are to first create the library (if it doesn't already exist), then to remove any old version of the particular object module (if this sequence has been used before), and finally add the new version of the module. The lib51 manual shows a single command, r e p l a c e, which should do the d e l e t e and a d d all in one step, but apparently that is only in later versions of that software.

Figure 6–8b   Assembler and Linker Commands

```
MATHTST:DO;

1   DLOAD:PROCEDURE (SIXTEEN) EXTERNAL;
         /* PUSH WORD ONTO A-TOP FILLS WITH 0S */
         DECLARE SIXTEEN WORD;
         END DLOAD;

    DADD:PROCEDURE EXTERNAL; /*A+B-> A (32 BIT)*/
         END DADD;

    DSUB:PROCEDURE EXTERNAL; /*A-B-> A (32 BIT)*/
         END DSUB;

    DMUL:PROCEDURE EXTERNAL; /*AxB-> A (16X16=32BIT)*/
         END DMUL;

    DDIV:PROCEDURE EXTERNAL; /*A/B-> A (32/16=16BIT)*/
         END DDIV;

    DRESULT:PROCEDURE WORD EXTERNAL;
         /* RETURN LOWER 16 BITS OF a */
         END DRESULT;
```

[1]These procedure declarations are necessary so the compiler knows how to prepare to use them. From the earlier tests with the code option, you know that PL/M will pass by location and return a word in r6,r7. The term E X T E R N A L is key to satisfying the compiler with *this* module, but, of course, without the assembly object modules on hand the linker will complain that it was unable to locate D L O A D, D A D D, and so on.

Figure 6–9a   PL/M—Assembly Linkage Example

```
  DECLARE  (A,B,C,D)WORD;
  A=999;
  B=1000;
2 CALL DLOAD(A);
  CALL DLOAD(A);
  CALL DMUL; /*999*999*/
  CALL DLOAD(B);
  CALL DLOAD(B);
  CALL DMUL; /*1000*1000*/
  CALL DSUB;
  C=DRESULT; /*1999*/
3 CALL DLOAD(B);
  CALL DLOAD(A);
  CALL DLOAD(A);
  CALL DMUL;
3 CALL DDIV;
  D=DRESULT; /*998*/
  END MATHTST;
```

[2]This example is only to test the use of large numbers to be sure the various carries and borrows are working. With the simulator working in hex notation, it was necessary to determine that 998=03e6, 999=03e7, 1000=03e8, and 999*999=0f3a71. A modern scientific calculator usually has these conversion functions built in.

[3]Because the assembly math has a stack (reverse Polish) architecture like many of the math chips and has no swapping on the stack, it is necessary to get **B** onto the stack *before* doing the **A * A** operation.

Figure 6–9a Continued

```
  plm51 mathtst.p51 debug
1 rl51 mathtst.obj,dimath.lib,plm51.lib ol
  oh mathtst
  copy mathtst.lst+mathtst.m51
```

[1]Notice that the linking includes the d i m a t h . l i b library which has the desired object file (and nothing else). It could have been done with d i m a t h . o b j instead in this particular case, but it is more general to use a library. The p l m 5 1 . l i b library is necessary for the initialization procedures included automatically in the module.

Figure 6–9b PL/M—Compile and Link Commands

```
   #define uint unsigned int
1  void dload(uint sixteen);
           /* push word onto stka—
              top fills with 0s */
   void dadd(void);
           /* a+b —> a (32 bit)*/
   void dsub(void);
           /* a-b —> a (32 bit)*/
   void dmul(void);
           /* axb -> a (16x16=32bit)*/
   void ddiv(void);
           /* a/b —> a (32/16=16bit)*/
   uint dresult(void);
           /* return lower 16 bits of A */

   void main(void){
       uint a,b,c,d;
       a=999;
       b=1000;
2      dload(a);
       dload(a);
       dmul(); /*999*999*/
       dload(b);
       dload(b);
       dmul(); /*1000*1000*/
       dsub();
       c=dresult(); /*1999*/
3      dload(b);
       dload(a);
       dload(a);
       dmul();
3      ddiv();
       d=dresult(); /*998*/
   }
```

[1]These function prototypes are necessary so the compiler knows how to prepare to use them. From the tests with the code option, you saw that the use of a l i e n will pass (like PL/M) by location and return a word in r 6 , r 7 . Without the assembly object modules on hand, the *linker* will complain that it was unable to locate d l o a d , d a d d, and so on.

[2]This example only tests the use of large numbers to be sure the various carries and borrows are working. With the simulator working in hex notation, it was necessary to determine that 998=03e6, 999=03e7, 1000=03e8, and 999*999=0f3a71. A modern scientific calculator usually has these conversion functions built in.

[3]Because the assembly math has a stack (reverse Polish) architecture like many of the math chips and has no swapping on the stack, it is necessary to get b onto the stack *before* doing the a * a operation.

Figure 6–10a   C—Assembly Linkage Example

```
    c51 mathtst.c debug
1   rl51 mathtst.obj,dimathc.lib,c51s.lib ol
    oh mathtst
    copy mathtst.lst+mathtst.m51
```

---

¹The **dimathc.lib** library was made to hold the revised dimath object file described in Fig. 6–8. Technically, *both* versions could be put in the same library with different names and the desired version could be picked by name. The **c51s.lib** library is supplied with the compiler and holds the entire set of functions that might be needed by the compiler including math, I/O, string and character, and initialization functions. The small library version is used because the default is small. The other choices are **compact** and **large**. The former is almost never used—only 256 bytes of off-chip RAM addressing is quite unusual unless it is I/O. The **large** model is more common, but in running the simulator and studying the code, it unnecessarily complicates things.

Figure 6–10b   C—Compile and Link Commands

impact. One example run with 4 different compilers, provided by Franklin—which is quite proud of their version 3 compiler, varied from 220 bytes plus library calls down to 34 bytes with no calls. This is probably a rapidly changing field and would need to be investigated carefully if efficiency is important.

There are some choices that you as the programmer can make that have significant impact on efficiency:

1. You can choose the small memory model (or **MAIN** vs. **AUXILIARY** for PL/M) where space allows to avoid the extensive use of the **movx** instruction.

2. With the large model you can give careful thought to which variables should be kept in data (MAIN) space because they are used often or serve as intermediate results.

3. You can consider the sequence of operations so as to finish with one variable before you work with the next one.

4. Minor things can help slightly also. For example, if you use a **for(;;)** loop, you can recognize that a **djnz** instruction is slightly more efficient than a **cjne** instruction and make the iterative loop count *down*.

5. A very immediate improvement, if your compiler is not particularly clever, is the use of shifting and rotating rather than multiplication and division. For example, a shift left by one causes a multiplication by 2.

6. Masking by logical ANDing is much more efficient than using a **MOD/%** operator. This is used in the buffering example of Chapter 9, for example.

7. Careful choice of data storage and array sizes based on the inherent binary nature of the computer can be a savings.

Obviously, these things may not make a large difference and you may need to experiment with the code option turned on to see what is best with your compiler. The more you can learn of assembly language and binary math operations, the more efficient you can become in your programming choices.

REVIEW AND BEYOND

1. What are reasons for using modular programming?
2. In each of the three languages, what is the scope of a variable defined at the top of a module without any special directives?
3. What must be done in both modules to enable a routine in one module to be called from another module?
4. What form of program normally goes into "libraries"? What happens when the library has more programs than the linker needs?

Chapter 7

8051 Family Hardware

Up to now the focus has been on *programming*. The examples have used only the most elementary aspects of the 8051's internal hardware. In addition to other internal hardware features, some discussion of adding external devices is needed.

This chapter first elaborates on some of the additional features of the hardware—the timers, the interrupts, and the normal serial port. It also discusses aspects of register banks and the stack.

A second section shows how program storage space is expanded and external RAM is added. Several schematics give a feel for the simplicity of expanding the memory and adding additional peripheral chips. (Appendix D surveys some commercially available boards in case you want to shortcut some of the hardware design process or reduce the amount of wiring to prototype a design. Techniques such as point-to-point wiring, wire-wrapping, and PC board layout are assumed to have come from other experience.)

A third section lists additional 8051 family members ("relatives") and describes a few of the additional features available. This coverage is of necessity quite brief, but it opens up the powerful features that make the family remain a viable option for many advanced requirements.

This is the chapter to study if you are expanding your horizons to rely more on the internal features of the 8051, are building up your own hardware, or are interested in how commercial board designs are put together.

TIMERS AND COUNTERS

The 8051 family devices have at least two internal timers. The 8052 relatives
have 3 timers, and some of the newer -500 and -700 numbered devices have various mixes of additional devices. But to start off, the two basic timers are t0 and
t1. They are 16-bit timer/counters with a variety of modes. If a timer is counting
the internal crystal-driven clock, it is in the *timer* mode. If it is counting transitions on a designated input pin of the 8051, it is in the *counter* mode. The choices
are made with the TMOD register (Fig. 7–1). The initial value in the timer is set by
two additional registers, TH1/TH0 and TL1/TL0. The counting or timing is
turned on with bits in TCON (Fig. 7–2). Both PL/M and C compilers provide
several **include** or **header** files which define all the registers of the particular
8051 family members.

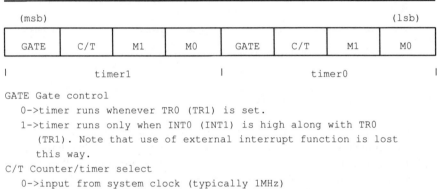

```
(msb)                                                               (lsb)

| GATE | C/T | M1 | M0 | GATE | C/T | M1 | M0 |

|         timer1              |          timer0                |
```

GATE Gate control
    0->timer runs whenever TR0 (TR1) is set.
    1->timer runs only when INT0 (INT1) is high along with TR0
       (TR1). Note that use of external interrupt function is lost
       this way.
C/T Counter/timer select
    0->input from system clock (typically 1MHz)
    1->input from TX0 (TX1) pin. Note that a count input must be
       high or low for at least 1 microsecond and have a maximum
       frequency no more than 500kHz.
MODE 00 13-bit counter lower 5 bits of TL0 (TL1) and all 8 bits
         of TH0 (TH1).
MODE 01 16-bit counter.
MODE 10 8-bit auto-reload. TH0 (TH1)->TL0 (TL1) when the
         latter overflows.
MODE 11(timer0) TL0 is 8-bit timer/counter controlled by timer0
         control bits TH0 is 8-bit timer (only) controlled by timer1
         control bits.
MODE 11(timer1) stops timer1 the same as setting TR1=0. (The
         most common use of mode 3 is to use timer1 as the baud rate
         generator and still have two 8-bit timers to generate
         interrupts.)

Figure 7–1 Timer Mode (TMOD) Register

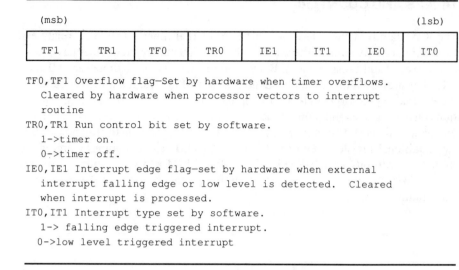

(msb)							(lsb)
TF1	TR1	TF0	TR0	IE1	IT1	IE0	IT0

```
TF0,TF1 Overflow flag—Set by hardware when timer overflows.
   Cleared by hardware when processor vectors to interrupt
   routine
TR0,TR1 Run control bit set by software.
   1->timer on.
   0->timer off.
IE0,IE1 Interrupt edge flag—set by hardware when external
   interrupt falling edge or low level is detected.  Cleared
   when interrupt is processed.
IT0,IT1 Interrupt type set by software.
   1-> falling edge triggered interrupt.
   0->low level triggered interrupt
```

Figure 7–2   Timer Control (TCON) Register

EXAMPLE: A 1MSEC TIMER

Figure 7–3 sets up t0 for 16-bit timing with a timeout after 1,000 clock cycles. It uses the **timer mode** where the system clock (typically a 12MHz internally divided by 12) is counted. For lower crystal frequencies, the timer will increment less often and some of the newer family members go up to 16MHz. In Fig. 7–3 it is assumed that a 12MHz crystal is used so a divide by 1000 gives a time of 1mSec. If an 11.059MHz crystal were used (common for proper function of the serial port discussed later), the divide to get 1mSec would be close to 922. That is 1000*11.059/12. Because the counter counts *up*, it is necessary to load the timer with a value which will reach ffff in 999 counts. Then the 1000th count will cause it to *roll over* to zero and cause the interrupt. To get a full count (65536 in the 16-bit mode), it is necessary to load zeros into the count registers.

FORGETTING THE TIMERS COUNT UP: To get a timer to overflow after 100 counts you must load –100 (or whatever that is in unsigned 2's complement).

```
#include <reg51.h>

2   TMOD=0x01;
1   TH0=~(1000/256);
    TL0=-(1000%256);
    TR0=1;
```

¹The divide / and mod % are used to obtain the byte splitting in C. Most likely the level of optimization will cause the compiler to compute the constant in advance and not actually use the routines from the library at run time. The expected expression for TH0 (–1000/256) did not yield the expected fc but rather fd (the 2's complement).

²The Franklin C compiler uses upper case for the register names and C is case sensitive here.

Figure 7–3a   Setting Up to Use Timer0—C

```
    mov tmod,#00000001b
1   mov th0,#0fch
    mov tl0,#18h
    setb tr0
```

¹The fc18 is –1000 in hex notation. With some assemblers it could have been written out as an expression and the assembler would compute the value of the constant for you.

Figure 7–3b   Setting Up to Use Timer0—ASM51

```
    $INCLUDE (REG51.DCL)
    TMOD=00000001B;
1   TH0=HIGH(-1000);
    TL0=LOW(-1000);
2   TR0=1;
```

¹The HIGH and LOW operators in PL/M avoid the problem of figuring out the high and low byte values for –1000.

²Rather than address the entire TCON register, the bit operations are used. Notice that these may vary between C compilers and are *not* ANSI standard.

Figure 7–3c   Setting Up to Use Timer0—PL/M

OTHER MODES

If you are using a timer in the **counter mode**, the appropriate bit of TMOD would be different. Such a change results in a pin on P3 becoming the counter input and the pin is lost for normal port activity. While discussing lost pins, it will be mentioned later that the upper 2 bits (6 and 7) of P3 are lost if external RAM is added (the RD and WR lines), bit 4 is lost for this (T0) counter input, and the other P3 bits could be lost if you are using serial I/O (bits 0 and 1), external interrupts (bits 2 and 3), or if the other timer serves as a counter (bit 5). Thus, P3 is usually not a *full* 8-bit port except for very simple single-chip applications.

For many ongoing counting and time-duration operations, it is best to use the timers as full-count devices. When the counter rolls over, it keeps on counting. If the value in the counter is read at the *start* of a count or time interval and then *subtracted* from the value at the *end* of the count or interval, the number obtained will be the number of counts in between, or the duration of the interval. Suppose a timer mode is being used to time the duration of a signal from a V-F (voltage to frequency) converter. If the counter value when a logic 1 arrives is 3754, and the count when the next logic 1 arrives is 4586, then the period of the V-F converter is 832 clock units or 832 microseconds (1.202kHz), with a 12MHz crystal (with an 11.059MHz crystal that count would equal about 903 microseconds or 1.107kHz). When the counter rolls over, there is no math problem as long as the readings are treated as 16-bit *unsigned* integers (WORD).

Another interesting mode is the **8-bit auto-reload mode**, which allows the timer to restart itself with a short count cycle. Every time the low byte, tl0, reaches 0, it is automatically reloaded with the value from the high byte, th0. Any program that needs a regular interrupt at a count period of 255 or less can avoid having to reload the counter in software at the start of the interrupt procedure.

TIMER2

If you have an 8052 or some of the other recent family additions, you have a *third* timer, T2, available for use. It is quite different to program and is much more flexible. T2 includes a **16-bit auto-reload mode** so that the counter/timer will automatically short cycle after it rolls over to 0 (like the 8-bit auto-reload mode on timers 0 and 1). The auto-reload can be initiated by a transition on an external pin, T2EX. This external pin can then be used for synchronization of the counter with some other hardware. It can also be a watchdog or timeout timer where, if a pulse doesn't arrive within the timer period, an interrupt can occur to indicate that the pulse is late or missing. The register for timer2 is described in Fig. 7-4.

T2 also has a **capture mode**, which transfers the instantaneous count value to another register pair for reading by the processor without the danger of the counter value rippling between the 2 bytes during the reading process. For rapidly

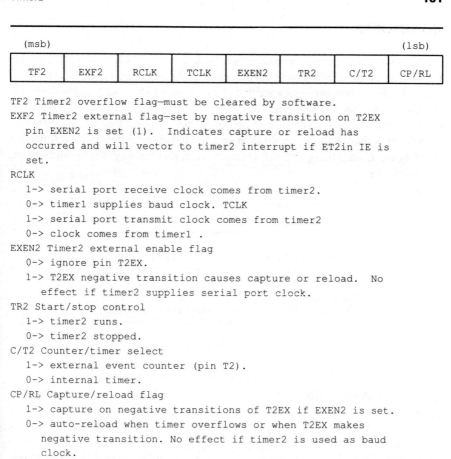

(msb)							(lsb)
TF2	EXF2	RCLK	TCLK	EXEN2	TR2	C/T2	CP/RL

TF2 Timer2 overflow flag—must be cleared by software.
EXF2 Timer2 external flag—set by negative transition on T2EX
   pin EXEN2 is set (1). Indicates capture or reload has
   occurred and will vector to timer2 interrupt if ET2in IE is
   set.
RCLK
   1-> serial port receive clock comes from timer2.
   0-> timer1 supplies baud clock. TCLK
   1-> serial port transmit clock comes from timer2
   0-> clock comes from timer1 .
EXEN2 Timer2 external enable flag
   0-> ignore pin T2EX.
   1-> T2EX negative transition causes capture or reload. No
      effect if timer2 supplies serial port clock.
TR2 Start/stop control
   1-> timer2 runs.
   0-> timer2 stopped.
C/T2 Counter/timer select
   1-> external event counter (pin T2).
   0-> internal timer.
CP/RL Capture/reload flag
   1-> capture on negative transitions of T2EX if EXEN2 is set.
   0-> auto-reload when timer overflows or when T2EX makes
      negative transition. No effect if timer2 is used as baud
      clock.

Figure 7–4   Timer2 Control (T2CON) Register

changing counts such as when an external pulse width or period is being measured with the internal clock, if the count is, say, 37ff when the *high* byte is read and has changed to 3800 by the time the *low* byte is read, the result will look like 3700. If the 37ff is *captured* instantaneously into another register pair, the computer can pick up the 37 and the ff at its leisure.

There are more details to the use of T2 as well as a proliferation of other additions on 8051 relatives. These additions vary significantly, although most of the relatives continue to support the basic 8051 functions (the '751 and '752 have only a single timer, but it has a 16-bit auto reload). There are numerous other registers for the various added features much as T2 added the T2CON register. The specific data sheets must be consulted to go further.

INTERRUPTS

Along with timers, vector interrupts are the key hardware components for real-time systems. As will be discussed at length in the following chapters, interrupts allow the development of software that does not need to be concerned with timing in other parts of the system. A math routine does not need to check the I/O devices every few instructions to be sure no port is needing service. Instead, the math can be written as though there is unlimited time available to do math and nothing else is going on. *If* something else does need service, this is recognized in the system by an interrupt.

Name of Interrupt	Vector Location	
EX0 EXTERNAL 0	(INTERRUPT 0)	0003H
ET0 TIMER 0	(INTERRUPT 1)	000BH
EX1 EXTERNAL 1	(INTERRUPT 2)	0013H
ET1 TIMER 1	(INTERRUPT 3)	001BH
ES SERIAL	(INTERRUPT 4)	0023H

Figure 7–5   8051 Family Hardware Interrupts

The 8051 interrupts are **vectored** in that, when an event happens to request the processor to stop what it is doing, the *hardware* changes the program counter (vectoring) to a specific value that depends on which interrupt occurred. The routine to which program flow vectors is called the **interrupt service routine**. For the 8051, the fixed vector addresses are shown in Fig. 7–5, so the compiler will put jumps at *those* addresses to the actual location of the service routines. As with most processors, this vectoring involves pushing the previous program counter value on to a stack (like a call), and the interrupt software (the interrupt service routine) ends much like any other routine (with a return). When interrupt vectoring takes place, the hardware disables all interrupts (the global E A bit of the I E register—Fig. 7–6). The flag bit that indicated either the external interrupt or the timer overflow is cleared by the hardware as the vectoring takes place. The flag bit for the serial port or timer2 (E S or E T 2) is not cleared by hardware. Usually you want to determine if R I or T I, or T F 2, or E X F 2 *caused* the interrupt. Your interrupt service routine should probably branch different ways depending on the source of the interrupt. You will have to be careful to clear those various flags *before* re-enabling the global interrupt (E A). If you don't, the flag will cause an immediate repeat of the interrupt! The 8051 actually has a specific return instruction, r e t i rather than r e t, for interrupts. It re-enables the system to recognize other interrupts. Thus, it is not necessary to reset E A in the normal use of interrupts—just enable the interrupts once during the program initialization.

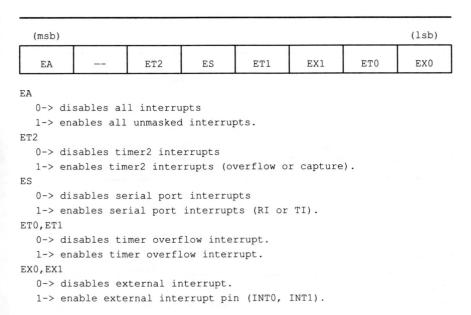

Figure 7–6 Interrupt Enable (IE) Register

MORE OR LESS INTERRUPT SOURCES

As Fig. 7–5 shows, there are at least five possible interrupts in an 8051 (six in an 8052, and up to 15 in other family members). A little more logic hardware could *OR* various hardware interrupt sources into one of the two external interrupt pins as long as the specific source that caused the interrupt can be determined by subsequent polling. This is not fully *vectored* interrupting, but it can be quite efficient where there are many interrupt sources that happen infrequently.

In general, the 8051 will be configured to use only some of the interrupts. The other interrupts will be **masked** so signals on those pins will not cause spurious interrupts and so the external pins involved can be used for their normal I/O port functions. If only one timer is in use to provide an interrupt, the other timer interrupt can be masked off. If a program is in a critical phase where it should ignore *all* interrupts, they can be disabled with E A. The interrupt enable register used for masking is shown in Fig. 7–6.

Because the 8051 directly supports only two levels of interrupt priority (see Fig. 7–7) along with the background noninterrupt level, *different* priorities should be used only when necessary.

When an interrupt is in progress, *no other interrupts* will be recognized of the same or lower priority. There are two priority levels for interrupts, set by the use of the IP register. Low-priority (priority 0) interrupt service routines are

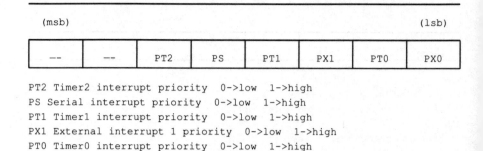

Figure 7-7   Interrupt Priority (IP) Register

masked when a high-priority or another low-priority routine is running. High-priority routines are masked when any other high-priority routine is running. Only when the r e t i is encountered or the E A is set does the interrupt hardware become sensitive to the other interrupts. Once set, the interrupt *request* flags will remain, whether they are set by hardware *or* software. It is possible to software-initiate an interrupt routine by setting T F 0, T F 1, I E 0, or I E 1 in T C O N; R I or T I in S C O N; or T F 2 or E X F 2 in T 2 C O N.

Likewise, it is possible to *clear* interrupt request flags in software before they are recognized by the vector interrupt hardware (usually while they are masked or blocked by a higher or equal priority interrupt in progress). Notice that an external *level-triggered* interrupt is not latched. If it is masked at the time it occurs and the level goes away before the interrupt is recognized, it will be totally ignored—the interrupt handler itself does not latch external level requests! You should be careful about long interrupt service routines if there are critical level-triggered external interrupt sources. The interplay of hardware and software in the use of interrupts can seem quite confusing at first, but the actual use is quite straightforward. Consult the "Architectural Overview" and the "Hardware Description" of the Intel data book to see these details from a different perspective. There is an interesting software/hardware method of simulating a third priority level described there.

REGISTER BANKS—CONTEXT SWITCHING

The 8051 has 4 **register banks** that are groups of 8 bytes at the start of internal memory. The designations r 0 . . . r 7 refer to those 8 bytes, depending on the setting of 2 bits in the P S W (Program Status Word). Those bits decide at any given moment whether references to r 0 - r 7 will go to RAM hex address 0–7, 8-a, 10–18, or 18–1f. Register banks allow very rapid **context switching**—the change typically involved in moving from one activity to another when an interrupt occurs.

Rather than the pushing and popping to a stack (normal with other processors), the change of two bits can save all eight registers. They will not be used until the bank select bits are restored and program flow returns to where it left off. In assembly, it is a matter of the programmer's choice, but, for linking with mixed language programming, the banks used in the assembly program can be specified so the linker will not allocate the bank as ordinary memory. In PL/M, the REGISTERBANK() directive makes the choice for the entire program module; the USING directive does the same for a specific PROCEDURE. In C, the choice of register banks depends on specific compiler directives (Franklin models its directives after PL/M).

High-priority interrupts can interrupt low-priority ones in progress, so it is necessary to pay attention to register banks. Unless you can be sure *no* use is made of r0 through r7 (because *you* wrote the assembly code), it is best to assign different register banks to routines of each priority. Incidentally, Franklin's C compiler can specify register-independent functions and code (by using only direct addresses). The current working register bank is set by two bits in PSW and can be controlled in PL/M and some C compilers with the *using* directive. Be sure that any routines used by the interrupt service routines also use the same register banks. Alternately, **reentrant** routines can be used, which can call themselves from themselves because they rely on the stack. The bulk of C library routines are usually reentrant, but you may need to check closely to see what happens to the registers when an interrupt comes. Also, external stacking routines may use the registers to load and unload values—don't blindly assume all the C compiler authors have considered *all* the implications of interrupts and register banks; some compilers are just adapted from more common *flat architecture* computers and may not have been tested in an interrupt environment.

For faster, more efficient code, use the interrupt service routines only for very simple operations; avoid using the longer variable types and math operations in interrupt routines. Instead, set flags so any lengthy processing can be done at the background level. For example, (1) bring raw serial characters into a buffer for line editing and command parsing, and (2) collect data and leave the averaging for later. This sort of design issue is an unavoidable part of true real-time systems.

THE 8051 UART

Most of the 8051 family of microcontrollers have on-board UARTs (Universal Asynchronous Receiver/Transmitter) for serial communication. If needed, level-shifting can be done to RS-232, in a separate chip (such as the MAXIM 232 or with the 1488/89 chip set) or to current loop with transistors. *Short* runs can use the TTL levels directly from the UART. The baud rate clock must be obtained from an internal timer—timer1 or, if available, timer2. If your application requires the full handshaking functions of RS-232, they must be handled in software using additional port pins. The registers for setting up the serial transmission are shown in Figs. 7–8 and 7–9.

(msb) (lsb)

SM0	SM1	SM2	REN	TB8	RB8	TI	RI

SM0,SM1

    00 -> mode 0   Serial data exits and enters RXD pin while TXD
    outputs the shift clock. 8 data bits with lsb first and shift
    rate of 1/12 of osc. freq. (about 1 MHz).

    01 -> mode 1   8-bit UART with baud rate determined by timer 1
    or timer 2. Transmission is 10-bits: start bit (0), 8 bits
    from sbuf (lsb first) and a stop bit (1). The choice of timer
    1 or timer 2 (only on 8052 members) is set by a bit in t2con.
    This is the normal UART mode.

    10 -> mode 2   9-bit UART with fixed baud rate. 11 bits
    transmitted/received including a 9th bit just before the stop
    bit. Baud rate is fosc/64 or fosc/32 depending on msb of
    pcon(see below). 9th bit transmitted comes from tb8 in scon.
    9th bit received goes into tb8 of scon.

    11 -> mode 3   9-bit UART with variable baud rate as with mode
    1. Same as mode 2 with respect to 9th bit.

SM2 Multiprocessor communication enable (modes 2,3)

    0 -> normal activation of RI when a character comes in.

    1 -> RI only enabled if bit 9 (RB8) comes in high

REN

    1 -> enables serial reception. In mode 0 it sets the shift-in
    mode rather than the shift-out mode.

    0 -> serial reception disabled. In mode 0 this sets the shift-
    out mode.

TB8   The ninth data bit to be sent in modes 2 or 3.

RB8   The ninth data bit that came in in modes 2 or 3. In mode 1
    it is the stop bit that was received.

TI   Transmit interrupt flag. Set at the end of the 8th bit
    (mode 0) or the start of the stop bit by the hardware to allow
    the software to know when to load the next outgoing character.
    Must be cleared by software.

RI   Receive interrupt flag. Set at the end of the 8th bit in
    mode 0 or half-way through the stop bit in any other mode
    (unless mode 2 or 3, sm2=1 and the RB8=0). Must be cleared by
    software.

Figure 7–8 Serial Port Control Register (SCON)

(msb)							(lsb)
SMOD	—	—	—	GF1	GF0	PD	IDL

SMOD Serial mode doubling bit
  0 -> modes 1,3 baud rate=timer 1 overflow rate/32; mode 2 baud
    rate fosc/64
  1 -> modes 1,3 baud rate=timer 1 overflow rate/16; mode 2 baud
    rate fosc/32
GF0,GF1 Two general purpose bits useful as flags
PD,IDL Power down bits useful only on CHMOS devices. See
  Intel's "Hardware Description of the 8051,8052 and 80C51" for
  more details on idle and power down modes.

Figure 7–9   Power Control Register (PCON)

EXAMPLE: SERIAL BUFFERING

The example of Fig. 7–10 shows a typical initialization of the 8051 serial port for transmission to a terminal or another computer. It shows initialization for 9600 baud, an interrupt routine to handle characters in both directions, and routines to buffer characters both ways. The background programs can "leave off" and "pick up" character strings in the buffers and the actual transfers to and from s bu f can be done by interrupts. When starting out, you may wish to ignore the buffer arrays and just focus on the serial port initialization.

The l o a d m s g routine loads the buffer array and signals the start of transmission. The buffers are each managed by two indices (in and out) and some flags. By making the buffers 32 bits long, the indices can be managed by simple logical A N D operations, which run more quickly than the M O D / % operations. When r b i n=r b o u t, then r b u f is *full*, and no more characters will be inserted. When t b i n=t b o u t, then t b u f is *empty*, and the transmit interrupt will be cleared to stop further requests by the UART. Other approaches to first-in-first-out buffers are possible—the focus here is on the serial interrupt routine. The assembly language version is omitted because the details would probably obscure the overall program approach.

```
 #include <reg51.h>
 #define uchar unsigned char
1 uchar xdata rbuf[32];
 uchar xdata tbuf[32];
 uchar rbin,rbout,tbin,tbout;
2 bit rfull,tempty,tdone;
 uchar code m1[]={"this is a test\r\n"};

3 void serint(void) interrupt 4 using 1{
    if (RI && ~rfull){
        rbuf[rbin]=SBUF;
        RI=0;
        rbin=++rbin & 0x1f;
        if (rbin==rbout) rfull=1;
    }
    else if (TI && ~tempty){
        SBUF=tbuf[tbout];
        TI=0;
        tbout=++tbout & 0x1f;
        if (tbout==tbin) tempty=1;
    }
    else if (TI){
        TI=0;
        tdone=1;
    }
 }

4 void loadmsg (uchar code *msgchar){
    while ((*msgchar!=0) &&
        ((((tbin+1)^tbout)& 0x1f)!=0)){
        /*test for buffer full*/
        tbuf[tbin]=*msgchar;
        msgchar++;
        tbin=++tbin & 0x1f;
```

[1]These are the buffers where background tasks leave off strings for transmission or pick up incoming strings.

[2]These pointers keep track of where in the buffers characters are going in or coming out.

[3]This is the serial port interface routine. Interrupt 4 is the RI or TI interrupt number. Note that it is necessary to test to see *which* signal caused the interrupt and that the flags hardware must be cleared by software. The full/empty flags keep characters from overflowing the buffers. The pointers are set up to use equality (which is quick to test) within the interrupt routine, while the "increment-and-fold-around" testing is done in the (noninterrupt) background. This sort of thinking is part of managing interrupt-driven activities.

[4]This function puts a character string into the transmit buffer. It retrieves only out of the code space—a separate function or a generic 3-byte pointer would be needed to pull strings out of xdata space.

Figure 7–10a   Serial Port Usage Example—C

```
5        if (tdone){
             TI=1; tempty=tdone=0;
             /*restart xmit if all finished*/
         }
     }
 }

6 void process (uchar ch){return;}
         /*who knows what?*/

  void processmsg(void){
     while (((rbout+1)^ rbin)!=0){
     /*not empty*/
     process(rbuf[rbout]);
     rbout=++rbout & 0x1f;
     }
  }
  void main(void){
     TMOD=0x20; /*timer 1 mode 2*/
     TH1=0xfd; /*9600 baud 11.o59mhZ*/
     TCON=0x40; /*start baud clock*/
     SCON=0x50; /*enable receive*/
     IE=0x90; /*enable serial int*/
     tempty=tdone=1;rfull=0;
     rbout=tbin=tbout=0;
     rbin=1; /*rbuf and tbuf empty*/
     for(;;){
7        loadmsg(&m1);
         processmsg();
     }
  }
```

---

[5]If transmission has previously finished, it is necessary to manually set T I to restart the interrupt routine for new transmissions.

[6]This null routine (just a return to simplify the simulation) could parse strings and make decisions based on incoming characters.

[7]The actual main program just loads the test message over and over and unloads any incoming characters. It is obviously not a final application.

Figure 7–10a Continued

```
    STEST:DO;
    $INCLUDE(REG51.DCL)
1   DECLARE RBUF(32)BYTE AUXILIARY;
    DECLARE TBUF(32)BYTE AUXILIARY;
2   DECLARE (RBIN,RBOUT,TBIN,TBOUT)BYTE;
    DECLARE (RFULL,TEMPTY,TDONE) BIT;
    DECLARE EOM LITERALLY '0';
    DECLARE M1(*)BYTE CONSTANT
                ('THIS IS A TEST',0DH,0AH,EOM);
3   SERINT:PROCEDURE INTERRUPT 4 USING 1;
    IF RI AND NOT(RFULL) THEN DO;
       RBUF(RBIN)=SBUF;
       RI=0;
       RBIN=(RBIN+1) AND 01FH;
       IF RBIN=RBOUT THEN RFULL=1;
    END;
    ELSE IF TI AND NOT(TEMPTY) THEN DO;
       SBUF=TBUF(TBOUT);
       TI=0;
       TBOUT=(TBOUT+1) AND 01FH;
       IF TBOUT=TBIN THEN TEMPTY=1;
    END;
    ELSE IF TI THEN DO;
       TI=0;
       TDONE=1;
    END;
    END SERINT;


4   LOADMSG:PROCEDURE (PTR);
    DECLARE PTR WORD MAIN;
    DECLARE (MSGCHAR BASED PTR)BYTE CONSTANT;
    DO WHILE (MSGCHAR<>EOM) AND
            ((((TBIN+1)XOR TBOUT)AND 01FH)<>0);
       /*TEST FOR BUFFER FULL*/
       TBUF(TBIN)=MSGCHAR;
       PTR=PTR+1;
       TBIN=(TBIN+1)AND 01FH;
```

[1]These are the buffers where background tasks leave off strings for transmission or pick up incoming strings.

[2]These pointers keep track of where in the buffers characters are going in or coming out.

[3]This is the serial port interface routine. Interrupt 4 is the RI or TI interrupt number. Note that it is necessary to test to see *which* signal caused the interrupt and that the flags hardware must be cleared by software. The *full/empty* flags keep characters from overflowing the buffers. The pointers are set up to use equality (which is quick to test) within the interrupt routine while the "increment-and-fold-around" testing is done in the (noninterrupt) background. This sort of thinking is part of managing interrupt-driven activities.

[4]This routine puts a character string into the transmit buffer. This routine retrieves only out of the CONSTANT space—a separate routine would be needed to pull strings out of AUXILIARY space.

Figure 7–10b   Serial Port Usage Example—PL/M

```
5       IF TDONE THEN DO;
            TI=1; TDONE=0; TEMPTY=0;
               /*RESTART XMIT IF ALL FINISHED*/
        END;
      END;
    END LOADMSG;

6 PROCESS:PROCEDURE (CHAR) /*EXTERNAL*/;
    DECLARE CHAR BYTE;
      /*WHO KNOWS WHAT?*/
      RETURN;
  END PROCESS;

  PROCESSMSG:PROCEDURE;
    DO WHILE ((RBOUT+1)XOR RBIN)<>0; /*NOT EMPTY*/
      CALL PROCESS(RBUF(RBOUT));
      RBOUT=(RBOUT+1)AND 01FH;
    END;
  END PROCESSMSG;
  TMOD=00100000B; /*TIMER 1 MODE 2*/
  TH1=0FDH; /*9600 BAUD 11.059MHz*/
  TCON=01000000B; /*START BAUD CLOCK*/
  SCON=01010000B; /*ENABLE RECEIVE*/
  IE=10010000B; /*ENABLE SERIAL INT*/
  TDONE,TEMPTY=1; RFULL=0;
  RBOUT,TBIN,TBOUT=0;
  RBIN=1; /*RBUF AND TBUF EMPTY*/

  DO WHILE 1;
7   CALL LOADMSG(.M1);
    CALL PROCESSMSG;
  END;
END STEST;
```

---

[5]If transmission has previously finished, it is necessary to manually set T I to restart the interrupt routine for new transmissions.

[6]This null routine (just a return to simplify the simulation) could parse strings and make decisions based on incoming characters.

[7]The actual main program just loads the test message over and over and unloads any incoming characters. It is obviously not a final application.

Figure 7–10b Continued

SHIFT REGISTER MODE

The serial port has two sometimes overlooked modes. There is a shift register mode, which is useful for simple I/O expansion, as well as communication and a ninth-bit mode, which is quite powerful for dedicated interconnection of processors. It should be mentioned that some chips also have an I^2C bus, which is described near the end of this chapter.

The **shift register mode** is useful for expanding ports with a minimum of hardware because a simple 8-bit shift register could be loaded directly for output or shifted in for input. Clocking at typically 1MHz, the 8 bits can be loaded in about 10 microseconds. Any number of shift registers can be cascaded for loading or reading through a series of writes to or reads from SBUF. This could be a very economical minimal system of I/O expansion if the ripple during the shifting is not important or if parallel-load latches are included with the shift registers.

A second use of the shift register mode is for communication between two processors. For short distances, the ability to communicate at 1MHz is quite impressive when compared with normal 9600 baud serial communication. Because the same port pins are involved for either direction, it is possible, with a few additional gates, to add communication among *several* processors. A bit or two from the normal parallel ports could set up connections and directions. Clearly there is no standard for such connections, and the system would be custom designed for each application.

NINTH-BIT MODE

An unusual feature of the 8051 serial port is the inclusion of a **ninth-bit** mode. This allows an extra bit in the serial transmission that is used to flag the receivers for special characters. For a simple network (among the choices discussed in Chapter 12), the ninth-bit scheme is quite interesting because the hardware allows receiving controllers to be interrupted only by characters having a one as an extra (ninth) bit. In that way, the transmitter can broadcast a byte with the ninth bit high as the *"Everybody pay attention"* byte. The byte (character) could hold the address of the node that should switch itself on and receive the following characters. All the following bytes (with the ninth bit low) could cause no interrupt to the other receivers, which had shut themselves off because their address was not sent. In this way, one processor could talk to a large number of other controllers without bothering controllers that were not addressed. To obtain a reply is not so simple with RS-232, because multiple transmitters on the same line would conflict. Open-collector schemes, or an arrangement with RS-422 or RS-485, are possibilities (as with BITBUS). From a networking sense, this would be quite primitive, with no collision detection, and such a system would have to operate in a strict master-slave fashion or with a token-passing arrangement. We go beyond the scope of this book, but such ground-up designs rapidly lead to very nonstandard systems.

For complete in-house developed systems, a variety of approaches are possible, but it is the lack of a *standard* that makes for much unnecessary diversity.

MEMORY EXPANSION FOR THE 8051

Probably the vast majority of the 8051 applications today use expanded memory for the code and often for the RAM also. The expansion is fairly straightforward—there are very few choices involved if you want the result to function properly! The schematic of Fig. 7–11 shows a simple single-chip expansion of the code space which is essential for *any* application using the 8031, which has no on-chip ROM or EPROM. Because of the continuing high price of the 8751, most applications not eventually intended for factory-masked ROM or demanding the single-chip feature use the off-chip EPROM configuration. The biggest drawback is the loss of at least two 8-bit ports to the addressing functions. Intel has had for many years a single-chip add-on chip, the 8157, that provides some EPROM and additional ports with an *internal* latch. For a minimum system, this could do the job with two chips. In these days, however, programs grow quite large and the more generic 27xxx EPROMS are a much better buy if the code size is large.

In any expansion of the 8051, it is necessary to account for the multiplexed address/data information on the lower 8 bits (P0). The low address must be latched when the A L E pulse goes low. Then the P S E N signal provides the memory read signal to the EPROM when the data should be driven on to the bus. The 74LS373 is the commonly used 8-bit latch for the lower byte of the address. The high byte of the address remains static on P 2 for the entire cycle and can be used directly for address decoding. In the given schematic, with only a single EPROM, additional decoding is unnecessary. (There is a rather seldom used expansion mode that gives only 8-bit addresses—pdata based on the assembly m o v x a , @ r 0 type commands.)

Once you have given up the ports, it is a small step to add external RAM. Again, the price of static RAM has continued to fall to the point where it is better to use generic RAM with a separate address latch even though Intel has the 8155 chip that provides the latch and ports along with the RAM. The more economical expansion of the 8051 is with the 8255 port chip, which is commonly available. Other chips to provide 8-bit latching for outputs or 8-bit buffering for inputs are available, but it is not certain that they would save in parts cost or board real estate. Figure 7–12 shows an expanded system with extra RAM and an 8255 to provide three 8-bit ports. In order to avoid an address decoder, the RAM is mapped at the base (0000) address and the 8255 is put up at the top address space. Perhaps a description of how to determine the addresses from the hardware is in order. Looking at the 6164 of Fig. 7–12, CS1/ is grounded, so it is always enabled. CS2 is tied to A13—since it is a true input, A13 must be high to select this chip. In binary, then the chip has RAM from 00100000 00000000 to 00111111 11111111 (0x2000 to 0x3fff). Since the 8255's CS is notted, it must be low to

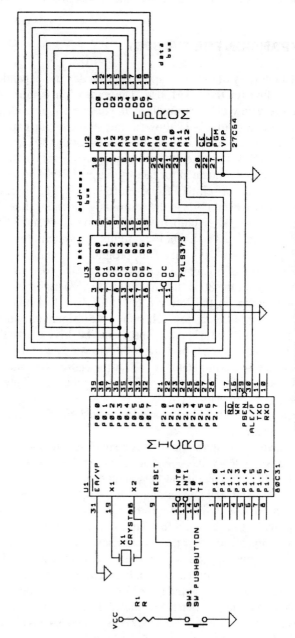

Figure 7-11 8051 with Expanded Code Memory

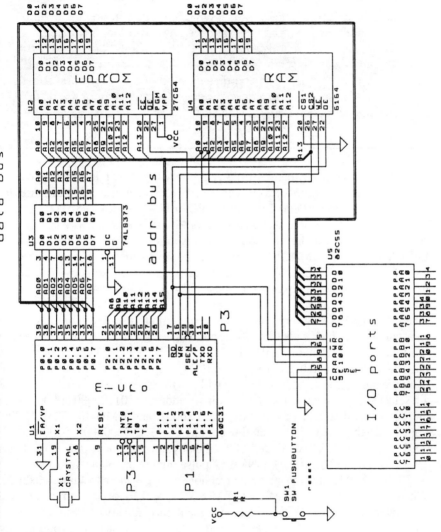

Figure 7-12 8051 with Expanded RAM, EPROM and Ports

ADDRESS DECODING SHORTCUTS: Although full address decoding with a 74138 is a good idea, for small systems it is possible to avoid the extra chip by using the individual upper address lines. For example, one chip can be selected when A15 is low, while another is selected when A14 is low. As long as you don't inadvertently address with *both* lines low, there is no problem. The potential problem is that you *can* address both devices at once which, if you are reading in, can cause bus contention. That probably won't damage the chips, but it can give strange results. A general suggestion is that you add a decoder only if you have more than 2 or 3 memory-mapped devices to select.

select the chip. Thus, anything that sets A13 low will suffice. It is normal to choose the other bits low, so, based on a quick look at the 8255 data sheet, the 4 addresses for the I/O chip are: PA=0x0000, PB=0x0001, PC=0x0002, and CMD=0x0003.

The RAM expansion takes up two bits of P3 for the RD/ and WR/ signals. The control for RAM (xdata or AUXILIARY) is thus different from the control for EPROM (code or CONSTANT) so they need not be decoded to different addresses.

Finally, an example of an added bus-interfaced peripheral device is shown in Fig. 7–13. The A-D chip is a common 8-bit device, which can be directly interfaced as shown. The status would be read in at one address and the data at another. A 74LS138 is a very common decoder chip to use if you are not taking shortcuts. Again, let's walk through the address decoding. First, the 8255 needs a low to its CS line. That means that the 74138 must set Y3 low, which requires 011 on CBA. That means that A15–A13 are 011, respectively. With 8K of RAM, there are only 3 bits of address to play with, so the other 3 select lines of the 138 are selected permanently. That gives a binary address of 01100000 000000xx for the 8255 (PA=0x6000, PB=0x6001, PC=0x6002, and CMD=0x6003). Unlike Fig. 7–12, the RAM is set at the bottom address, 00000000 00000000 to 00011111 11111111 (0x0000 to 0x1fff). Finally, the A-D took some work. It is a 10-bit converter. To get the CONVERT pulse high requires the WR/ low while A15–A13 are 001, thus, to start a conversion, write *anything* to address 0x2000. When the conversion is done the DATAREADY/ will go low, which can be used as an interrupt (negative edge triggered as mentioned already), or you can simply poll the pin as P3.2 or a bit at address 0xB2. Once the data is ready, you can pick up the lower 8 bits from address 0x4000 and the high two bits as the bottom 2 bits in from address 0x2000. This may not be the friendliest to program but, once you have decoded it and set up the #defines at the top, you can still use them in a convenient manner. As an added exercise, check the addresses of the devices in Fig. 5–5a and Fig. 10–1a.

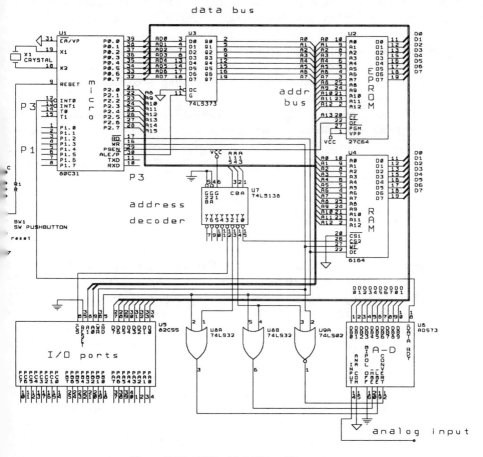

Figure 7–13   8051 with Additional Expansion

COMMERCIAL BOARDS

Although it is quite possible to wire-wrap an 8051 prototype system, it is probably more economical for commercial project development to make use of prebuilt
boards on the market. Most of them provide all the features described above for
port and memory expansion and usually come with a monitor program to
download code from a host (PC) to RAM by way of a serial port. Most of the
examples in this book were first tested on a Micromint board used in the student
laboratory.

The drawbacks of the commercial boards are related to the use of a monitor—the resources to communicate to the host development PC as well as the
code and RAM space taken by the monitor are unavailable to your application

program during development. The economics of a monitor are usually better than using an In-Circuit Emulator, but you have to decide what the cost goals for the final product are as well as the funds available for development tools. The commercial boards have the advantage of arriving already tested with some degree of documentation. There are apparently products with directly connecting add-on boards for A-D, LCD/keyboard interface, and the like. Such hardware could save significant design and debugging time!

The list of boards in Appendix C is probably not complete and may be obsolete within a few years as companies and designs come and go, but it provides some idea of the *sorts* of products available. Because the 8051 design is fairly mature, the only expected changes are designs that incorporate the newer members of the 8051 family.

THE 8051 FAMILY MEMBERS

The family of 8051 relatives seems to be growing steadily, to the point where the core is even available for ASIC design libraries. The processor can be put on-chip with other circuitry to create a truly custom device. Most applications probably don't have the volume to justify such cost yet, but automated design tools may be changing that.

The table that follows attempts to list the known 8051 relatives. Most members have versions without on-chip factory-masked ROM and a few have EPROM on-chip. Factory-masked versions are generally available (the usual distinction is the presence of a 3 instead of a 5 where no ROM is present and a 7 where EPROM is present). Masked versions would not be used for project *development* unless the EA (external access) pin is set to disable its use. The blank spaces in the table only indicate that specific data were not located. This table is only a rough attempt to give an overview of the capabilities available. The specific data sheets should be studied for more details. For example, the FA ... FC versions have a host of modes and up to 5 outputs from the fourth timer, which makes them quite flexible, but which is too complex to describe in a table like this. Major manufacturers, besides Intel, are Signetics and Siemens.

SPECIAL TIMER FEATURES IN THE 8051 EXPANDED FAMILY

Although it is not intended to cover *all* the details of the various 8051 relatives, a few features are different enough to merit more description to help you decide if they are useful.

Pulse-width modulated (PWM) outputs are quite useful in driving DC motors at variable speed without getting into analog power circuitry. The inductance of the motor must be taken into account as well as the possibility of inductive spikes and noise, but the method basically runs the motor with a series of constant-frequency pulses, the wider the pulses, the faster the motor turns.

Pulse-width modulated (PWM) outputs can also provide analog outputs

8051 Relatives

Part Number		Rom	Ram	Timers	Parallel Ports		Other Features
'31/'51		4K	128	2	4	8-bit	UART original
'31AH/'51H		4k	128	2	4	8-bit	UART, HMOS
'L31/'L51		4k	128	2	4	8-bit	UART, low voltage, low pwr.
'C51/'C31	[2]	4K	128	2	4	8-bit	UART
'C51FA	[2]	8K	256	4[1]	4	8-bit	UART
'C51FB	[2]	16k	256	4[1]	4	8-bit	UART
'C51FC	[2]	32K	256	4[1]	4	8-bit	UART
'52/'32		8K	256	3	4	8-bit	UART, encryption bits
'C32	[all]	8K	256	3	4	8-bit	UART
'53AH	[2]	8K	256	2	4	8-bit	UART
'053	[3]	8K	192	2	3	8-bit	TV display PWM,D-A
					1	4-bit	
'054	[3]	16K	192	2	3	8-bit	TV display PWM,D-A
					1	4-bit	
'44	[2]	4K	196	2	4	8-bit	SDLC controller
'352	[4]	32K	256	3	4	8-bit	UART, factory mask only
'CL410	[3]	4k	128	2	4	8-bit	I2C, 1.5v operation
'C451	[3]	4K	128	2	7	8-bit[3]	
'C452	[3]						2 DMA,FIFO
'513	[4]	16K	256	3			UART, factory mask only
'515/'535	[4]	8/16K	256	3+WD	6	8-bit	UART,8-bit(8ch)A-D, 4chPWM
'C515/'C5 35	[4]	8K	256	3+WD	7	8-bit	UART,8-bit(8ch)A-D,4chPWM
'C517/'C537	[4]	8K	256	4+2WD	8	8-bit	USART,8-bit(12ch)A-D,
					1	4-bit	21chPWM,UART,math unit
'C517A/'C537A	[4]	32K	256	4+2WD	8	8-bit	USART,8-bit(12ch)A-D,
					1	4-bit	21chPWM,UART,math unit
'524	[3]	16k	512	3+WD	4	8-bit	UART,I2C
'528	[3]	32k	512	3+WD	4	8-bit	UART,I2C
'C537	[4]			4	6	8-bit	UART(2),32-bit math
'C541		16K	256	2+WD	4	8-bit	
'C552	[3]	8k	256	4[1]	6	8-bit	UART,I2C,10-bit(8ch)A/D, 8 high-speed outputs, 4 cap.regs,2 PWM output
'C562	[3]	8k	256	3+WD	6	8-bit	UART,8-bit(8ch)A/D,PWM+ capture counters
'C592	[3]	16k	512	2+WD	6	8-bit	UART,10-bit(8ch)A/D, CAN bus
'C652	[3]	8k	256	4	4	8-bit	UART,I2C
'C654	[3]	16k	256	4	4	8-bit	UART,I2C
'C751	[3]	2k	64	1	2	8-bit	I2C,small pkg.
					1	3-bit	
'C752	[3]	2k 64	64	1	2	8-bit	I2C,8-bit(5ch)A/D
					1	5-bit	1 PWM output
'C851	[3]	4k	128	2	4	8-bit	256 byte EEPROM
'C852	[3]	6K	256	2	1	2-bit	2K EEPROM

[1] Includes up/down, watchdog, and pulsewidth modes.
[2] Intel only
[3] Philips/Signetics
[4] Siemens

with the addition of an integrating circuit, which can be as simple as a resistor and a capacitor. With the '552, for example, the PWM outputs are controlled by three registers. One determines the repetition rate for both outputs and the other two registers determine the width of the respective outputs. The width is broken up into a resolution of 255 and the repetition rate can be from 92Hz to about 23kHz.

The watchdog timer is used to reset the microcontroller if it enters an error state (perhaps through noise or programming errors). Basically, your program must get back to restart the timeout before the time expires or the timer will cause a reset of the processor. It requires that you determine how long you can allow the processor to be "lost" and how often you can get around to restarting the timer. There is a complicated enough reloading process to make it unlikely that random code would restart the timer. It can be enabled by tying a pin low and cannot be disabled in software. Particularly if your system has EEPROM, it may be desirable to recognize the difference between a "cold" and "warm" start to enable a smoother restart from where things left off.

High-speed capture and compare logic is useful where precise timing is needed beyond that possible with program flow. External edges can cause the value of an internal counter to be internally latched by hardware. This makes it possible to measure time intervals much more accurately. The compare logic can cause timed interrupts and also provide precisely timed outputs somewhat like the pulse width modulated outputs.

SPECIAL SERIAL COMMUNICATION FEATURES

In addition to the little-known shift-register and ninth-bit modes of the standard 8051, there are two other serial options that deserve attention. First is the SDLC controller in the '44, and second is the I^2C bus of the '552, '751, and '752.

It seems odd to describe the '44 as a special member of the family because it dates back almost to the start of the 8051. (Some of you may have heard of the 8041 and 8042, which were the parallel-port relatives not based on the 8051 core.) But the '44 has been the heart of the BITBUS protocol, which is discussed in Chapter 12. Basically, it is an 8051 with the UART replaced with a SDLC (IBM's Synchronous Data Link Control protocol) controller. The serial interface unit handles all the details of serial communication: zero bit insertion/deletion, address recognition, CRC (Cyclic Redundancy Check), and frame sequence check are all done automatically. By handling all the communication, the serial interface unit frees the CPU to concentrate on other tasks. When a message arrives or is sent, the transfer occurs in a shared memory area above address 128. With BITBUS, this communication is the heart of *distributed* control. The protocol (not discussed here) is well defined and allows lengthy blocks of data to be sent with addresses, length indication, and an error-checking code (somewhat like a parity bit on a byte). The SDLC protocol requires one node to be the master and the rest to function as slaves—slaves speak only when spoken to. That avoids the problem of contention—two devices talking at once—but it means that slaves must be polled by the master and messages to other slaves have to be relayed by the master. The protocol is quite fast (up to 2.4MHz) and has a good degree of error checking, so it is a distinct improvement over simple RS-232 communication.

A second, much newer serial feature is the I²C bus developed by Signetics/Philips. This bus is an open-collector, wired-or bus, which allows multiple masters with a resolution of bus contention. It is considerably slower (up to about 100kHz), but it is adaptable to much slower devices. It involves two lines, SCL and SDA—the clock and data, respectively. Without getting into details, data transfer occurs when the master of the moment drives the clock line. If it is sending then it also drives the data line. Otherwise, the replying slave drives the data line in synchronism with the clock from the master. If the master is clocking too fast, the slave can hold the clock low to make it wait. That is where the wired-or feature comes in handy. There are details relating to start and stop conditions as well as arbitration when several devices attempt to become the master at the same time.

The I²C protocol can be generated with normal port lines on processors, which do not have direct support. The data rate is slow enough to allow it to be tracked in software. Aside from communication between processors, the protocol is becoming more attractive as additional peripheral devices become available. In addition to small RAM and EEPROMs, there are D-A converters, port expanders, LCD drivers, voice and frequency synthesizers, and several radio and telecommunication devices that all respond to the I²C bus.

REVIEW AND BEYOND

1. How many priorities are inherent in the 8051 interrupts?
2. What is the difference between a timer mode and a counter mode?
3. What provides the timing for serial communication with the 8051 family?
4. Why does the program example initialize the counter with a "negative" value? Could this be avoided? Explain.
5. What are the major "costs" of expanding the RAM of the 8051?
6. Why would one possibly use the other members of the 8051 family? What features stand out in your mind?
7. Why would you use a commercial board with an 8051 rather than wiring up your own? When would you not do so? Why?

Chapter 8

Real-Time Ideas

BEYOND SINGLE-PROGRAM THINKING

At this point, I have talked about traditional areas of microprocessor programming relying mostly on examples using parallel ports. This is adequate in many cases, but when the applications grow more involved and time-critical, the single-program approach becomes awkward to handle. Particularly when there are several external hardware devices to serve at the same time, the challenge is to make sure all the hardware is satisfied. This chapter introduces the basic ideas of multitasking as well as six commercial multitasking operating systems for the 8051. Chapters 9–11 go into more detail on timing, communication, and interrupts. Each chapter begins with an example of the write-it-yourself method and then sketches how you would do the same things using *each* of the commercial multitasking operating systems. Chapter 12 very briefly goes into interconnected multiple controllers sometimes called distributed control.

WHAT IS "REAL TIME?"

Although the term means different things to different people, we will apply **real time** to any system that responds to inputs and supply outputs *fast enough* to meet user or external hardware requirements. For example, a keyboard entry system is real time if it supplies quick enough feedback for you to feel confident that the system "heard" you. If a "beep" is fed back to you within 100mSec, you feel the system recognized the input "right away." So, a keyscan routine that repeated every 100mSec would probably meet the requirement of *fast enough*. Likewise, you cannot assimilate new digital display information more quickly than perhaps 5 times a second, so numeric values that are rapidly changing might just as well be updated only a few times a second. On the other hand, a stepper motor ought to be sent a new step pulse at a regular time with millisecond precision, so this is a much more critical time requirement. A 9600 baud serial port probably has a single character buffer, so every *incoming* character must be picked up within about 1mSec (10 bits @9600 bits/sec = 960 char/sec). *Outgoing* (asynchronous) characters can go whenever the processor is not busy, assuming there is no other time constraint on message transfer. The same considerations would apply to collecting data via an A-D converter. Depending on how rapidly the incoming voltage is changing, the reading might need immediate retrieval or might be held for quite some time. *How rapidly should a flow valve be adjusted for a process? How quickly should a motor be supplied a new speed setting?* All of these are questions relating to what is *fast enough*.

Real-time systems differ sharply from the sorts of systems that handle data processing. The traditional large-computer system is *not* real time. A batch of input data is taken from a file, and results are stored to a file or sent to a printer some time later. Whatever the data and whatever the processing, there is some time delay from the request for processing to the obtaining of the results. Because there is some time delay involved, it isn't real time. If the delay were less, presumably the user would be more effective. A real-time system is one that responds so quickly that the users and hardware would not be any more effective if the response were faster. If the data processing job were done in 20 seconds, perhaps it would be better if it were done in 1 second. But if it were finished 100mSec from when the operator pushed the last key, it is doubtful that speeding it up to 1mSec could even be perceptible to a human.

Some areas where there is a push for real time include speech recognition, image processing, and digital signal processing. Imagine a dynamic display of the spectral energy distribution of a speech signal as it is picked up by a microphone rather than later from a recording! Or consider the possibilities of issuing canceling signals to make an airplane invisible to radar. How about a robotic system that can recognize objects as they travel down a conveyor line and sort them on the way past? Indeed, much of the *high-tech* focus of the microcomputer field relates to these very compute-intensive applications. The part of the real-time application world covered in this book is (fortunately) much less demanding. Some of the high-tech applications could never work in real time on an 8051, but it is important to realize that the difference between the big, fast applications and 8051 jobs is more one of scale than of technique. Big-system data-handling programmers might still view the techniques discussed here as advanced, but that is because hardware-related embedded systems are in quite a different field.

TASKS

For the discussion that follows, we also need to define a **task**. In its simplest form, a task is any *one* thing to be done by a controller. Think of an army with an overall mission to accomplish and many individual soldiers to carry it out. The overall mission, *"Drive out that dictator,"* is the **job**, but the tasks might be, *"Drive this supply truck to the front lines,"* or, *"Sit in this trench until ordered to advance."* For a controller, a task might be to wait for and process input from a keyboard or to keep a motor running at a set speed. A task can be as simple as a single routine having a few lines of code, or it can be an entire set of nested routines. The task divisions can be arbitrary as long as they relate to the functional parts of the overall job. Most important, tasks represent a *way of thinking* that logically divides the job and leads to the effective use of the microcontroller.

MULTITASKING

Getting the computer to manage several tasks seemingly simultaneously is the heart of **multitasking**. You may write tasks separately as though nothing else ever happens—they usually are written as separate modules in the style of Chapter 6. The ideas involved are not difficult to visualize if you compare them with everyday activities, but they differ from those of normal programming. You should develop the frame of mind that says, *"If it isn't ready yet, go on. I'll come back to it later. Right now there may be something else to do."* If the tasks are interrelated, you will have to plan signaling between them. Once you've understood the ideas of the next few chapters, it will be easy to move to complicated systems involving many calls with many parameters. Real-time systems are becoming increasingly important as the dedicated microcontroller becomes more and more prevalent. Knowing the *concepts*, you can have a role in the *implementation*.

Several different approaches to multitasking need consideration. One school of thought identifies multitasking with what might better be called "multiple tasking." It is **round-robin** philosophy where there are indeed tasks, but they run one after the other in turn, without the ideas of preemption or priority that are essential for external real-time events. Although this approach may include interrupt service routines, attention is centered on what might be called the *background* tasks. In such a scheme, each task waits its turn to run. If a task has nothing to do at the time, it immediately passes control to the next task, but if it has a lot to do, it takes all the time it needs to finish. Obviously, such a system avoids the overhead of an operating system, but such a scheme requires your close attention to avoid long latency in getting around the loop. You have to be sure to break big tasks into several small tasks or make intermediate pause points to be sure the running task doesn't hog the processor.

The **time slice** approach addresses some of these problems. It provides protection from hogging the processor by arbitrarily switching out long-running tasks at regular intervals. There must be a timer-driven interrupt marking off time slices so control of timing is removed from the running task. This approach is good for data processing applications and was the heart of time-shared computing systems, which were common before the personal computer revolution. In its basic form, though, it overlooks the idea of *preemption* needed for nondeterministic systems—ones where urgent tasks may unexpectedly or unpredictably need immediate attention.

A **scheduler** is somewhat related to a time slice system in that it keeps track of time for the various tasks. Typically, however, there are some tasks that are short and repetitive such as scanning external inputs or sending regular outputs. These short tasks may come frequently or only occasionally and at different intervals. Any leftover processor time is used for a background task, which is often a much less urgent data processing or decision-making activity.

Finally, **priority-based preemptive multitasking** systems switch tasks based on events. They can be periodic events like a scheduler, or dynamically changing events based on outside inputs. An **event** is anything that might cause the system to change tasks. It could be an *external* interrupt, an *internal* signal or message sent from another task, or the expiration of a waiting period. The primary point is, events can lead to a change in which task is running.

REAL-TIME HARDWARE REQUIREMENTS

The 8051 microcontroller has two internal hardware features that are virtually essential for solving real-time problems: timers and interrupts. Although both are covered in the data sheets and application notes from the manufacturers, some details were included in Chapter 7 because they relate so directly to multitasking.

Internal **timers** allow tasks to go out at regular intervals to scan inputs or change outputs. With many tasks to be done, managing time delays with a separate timer instead of a software delay loop allows other tasks to run while a given task is waiting.

Coupled with timers, a real-time system depends on **interrupts**. In addition to the externally triggered interrupts you may have studied elsewhere, it is crucial to have a real-time interrupt so the operating system can come in periodically in the middle of the tasks without requiring any special care in the programming of the tasks. Most multitasking systems keep track of time for all the tasks that are time dependent. Of course, multitasking can also include task switching based on other interrupts, but this will be discussed further in a later chapter.

WHY AN OPERATING SYSTEM FOR MULTITASKING?

Before going on, perhaps a few general words on *why* you might want to use a multitasking operating system are in order. First, multitasking helps you avoid some of the programming headaches that come with developing embedded control applications. The system calls formalize and standardize things, making it easier to include multitasking. The programs will be more readable to someone else and the interactions will be more consistently defined.

Second, unlike other multitasking systems, *some* of the 8051 operating systems are very small and control-oriented. They have been designed to handle multiple real-time inputs and outputs very quickly and efficiently. You do not give up as much as might be expected by using one instead of writing your own. On the other hand, some of the systems are adapted from ones for other processors that are more stack-oriented, and there are a variety of approaches to context switching (discussed later). The size and speed requirements of your application are important in the choice of a system.

Finally, if your projects grow and change as much as mine have, you will not be forced to reconsider and scrap large pieces as the "creeping features" come along. The task interactions will not be as buried in the subtle, clever (and obscure!) approach you carefully developed from scratch at the beginning. At the same time, programming in C should be fairly portable to some other higher-performance processor if necessary. In the debate *between* operating systems, it will be pointed out that "pure" ANSI C should be easier to port to a different processor, so you have to weigh speed and efficiency now against code portability in the future. There is no question that C is preferable to assembly if any change is anticipated.

HOW DO OPERATING SYSTEMS REALLY WORK?

Multitasking is not magic! It is simply an approach that has been developed to formalize programming that deals with multiple (usually real-time) activities. Simple switch inputs and display outputs are just as much multitasking as advanced personal computers running jobs for multiple users at the "same" time. The key is in a means to frequently check (usually) in the middle of a task for *other* tasks that have become more urgent than the *current* activity. A clever programmer can do that without any additional hardware or software, but it becomes very complicated to manage as programs grow larger, and the chance for errors increases when *several* programmers are developing pieces of the overall program.

Chapters 9, 10, and 11 will go over various aspects of multitasking. Along with direct-programmed multitasking approaches, the chapters will each describe features of several commercial *interrupt-driven, priority-based* multitasking operating systems. A whole book would be needed to fully illustrate the practical uses of such systems, but I will attempt to give you a *feel* for how these operating systems handle real-time multitasking. Appendix E lists the system calls for each system by category.

Keep in mind that multitasking operating systems are really just additional program code, which runs every so often to keep track of tasks. Usually *one* high-priority task *is* the operating system (although it may not be thought of as a task). It comes into play (1) any time a running task issues a system call, (2) at each time increment (system tick—usually an internal timer-caused interrupt), and, depending on the operating system, (3) at any other interrupt. At such times the program flow returns to the operating system program, which "wakes up," recognizes what has happened, and saves the return address pointer of the program that *was* running (found on the on-chip stack). The operating system program *then* adjusts tables that keep track of task state, interval count, timeout count, and so on. Having updated these tables, the operating system program *then* searches through various tables and queues to determine which task should now be made the running task. *Finally*, task switching, if required, is accomplished by at least *replacing* the return address pointer in the microcomputer's stack (normally a dangerous practice!) and issuing a r e t i instruction.

Incidentally, one of the big differences between various 8051 operating systems is the way they handle **context switching**. The context is all the values being used temporarily by a task, which should not be changed in the middle of the task. It might only involve the accumulator, the B register, and the eight other registers, or it might include the stack, various other on-chip variables, and even some artificial off-chip stack. The small "lean" operating systems rely on the programmer to avoid possible conflicts by using fixed, separate addresses for each task's parameters, using the register bank feature, and avoiding many different priority levels. Then there isn't much about one task that would be overwritten by another task. The bigger systems allow almost anything in the way of task memory usage and "clean the slate" every time a new task is started by saving almost everything out to off-chip RAM to be safe. One even copies the *entire* on-chip RAM to off-chip RAM at each context switch! The programmer then has less restrictions, but as the code becomes larger the system cannot change tasks as quickly.

From the perspective of the individual tasks you write, a system call is simply a call to a routine, which, after some period of time, "finishes," and your task's program flow resumes. The fact that there are *other* tasks is invisible to the program flow.

DCX51

Intel's DCX (Distributed Control eXecutive) is one example of a multitasking operating system (certainly the first commercial OS for the 8051). It is among the smallest and simplest of them all. There are four predefined events for DCX: (1) *Interrupt* events coming directly from hardware interrupts on the 8051, (2) *Message* events, which are software-initiated by other tasks sending a message to a given task, (3) *Interval* events that are regularly occurring events at some multiple of the system clock (usually 1mSec), and (4) *Timeout* events that come if the other desired event(s) doesn't take place within a specified time after a task begins waiting.

Tasks can be in one of three possible task states, as shown in Fig. 8–1. *Running* is the state of a task that is being executed by the processor at the moment. Only one task can be running at a time. *Asleep* is the state of a task that does not need to run until one of the events occurs. *Ready* is the state of a task that *would* run *if* something of higher priority were not running at the moment. The

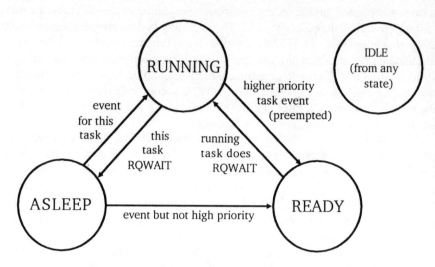

Figure 8-1 Task States

ready tasks queue up for their turn to run. Finally, there are the *ready/preempted* tasks that *were* running when something of higher priority "woke up" and took over the processor. A preempted task is the first to resume running when its priority level again has use of the processor. This happens when the higher priority task finishes and "goes to sleep."

DCX is first of all the software base for BITBUS systems (see Chapter 12) where it is put on an 8044 chip, but it can be linked in with application code on an 8051 when the automatic off-chip message features are abandoned. The operating system and a communication task fit within the 4K of on-chip ROM, but the system claims the vast majority of on-chip RAM for stack and message buffers as well as 300 bytes of off-chip RAM for its various tables and queues. DCX tasks have to keep most of their variables in off-chip RAM. There are 12 system calls in release 2 of DCX.

DCX is virtually nonconfigurable; in making it small, a number of choices were eliminated. There can be only 8 tasks on record at a time (to make for short tables and queues), there are only 4 kinds of events, each task has only one mailbox, and there are no signaling mechanisms directly equivalent to semaphores. For context switching, the system uses the inherent register-bank structure. This makes switching quite fast but limits to three priorities. By assigning separate pieces of internal stack for each task, the system leaves no reasonable space for the normally stack-oriented characteristics of C. Fortunately, the better C compilers for the 8051 avoid the internal stack at the cost of reentrancy and recursion. Although these restrictions may bother large-system-oriented programmers, they do makes it quite well adapted to the constraints of the 8051 family. DCX is supplied as an object file, which must be linked in with your application programs, or, if using BITBUS, DCX is supplied as firmware factory masked in the 8044 chip (see Chapter 12).

DCE51

A very close relative, functionally, of DCX is the DCE system from Iota Systems and MDS Technologies. It uses the *same* system call names as DCX with a few additions to add more memory management for off-chip RAM giving 19 system calls in all. Unlike DCX, it runs *all* tasks in register bank 0 and uses bank 1 for itself. This allows more priorities, but the system limits itself to five levels. The operating system context switch then involves writing the register bank off to off-chip RAM if another task is to resume. The DCE literature seems to focus on assembly language application examples. The operating system is supplied in an object file, which you link with the application programs and with a "configuration file," which you modify and assemble. Particularly interesting developments are a "monitor" system and a distributed computer system, which together approach the BITBUS tools without requiring the 8044 chip (they rely on the standard serial port of the normal 8051 family members).

RTXTINY

Another operating system, from Franklin Software, is called RTX. It is quite new, but appears to concentrate on the very small applications much like DCX and DCE. In addition to the **RTXtiny** version, which runs *totally* in on-chip RAM, there is an RTX51, which more closely resembles DCX and DCE. RTXtiny uses no more than 64 bytes of RAM (depending on how many of the 16 possible tasks are defined), has code of only about 800 bytes, and has only 6 system calls. RTX51 is still modest sized, requires some 650 bytes of off-chip RAM, up to 46 bytes of on-chip RAM, and 6K–8K of code space. RTX has 17 system calls, supports 19 active tasks at one time, and includes message passing as well as timing, interrupts, task signaling, and memory pool management. RTXtiny is a "subset" supporting only timing, interrupts, and intertask signaling, which is enough to build up virtually any application. Both systems will run tasks in round-robin fashion, but the RTX (large) version is similar to DCX and DCE in providing priority levels for tasks. RTX51 time slices *equal*-priority tasks whereas DCX and DCX let equal-priority tasks run to a wait on a first-come-first-served basis.

Because the available information was particularly sparse at the time of writing, the examples that use RTX will all use only RTXtiny. This will be efficient, but will become more difficult when messages are involved.

TWO BASIC GROUPS OF RTOS

The three *small* systems, DCX, DCE, and RTX differ slightly but they all represent the very small, possibly restricting approach, which is probably superior for small, fast systems tailored specifically to the 8051. The remaining four systems are quite different in that they are apparently ported over to the 8051 from other processors. They appear to have been originally written in C for some other processor and then adapted to the 8051's unusual architecture with some special context switching techniques and, in some cases, an assembly language piece for the most critical parts. In preserving some of the features of the operating system from the other processors, efficiency and speed may have suffered on the 8051, but the range of system calls and the high degree of flexibility may outweigh those drawbacks. You will need to be guided by the size and complexity of your applications in making a choice.

USX

United States Software Corporation markets a multitasking operating system it names *MULTITASK!*, but which will be designated *USX*. It is a considerably more complex system than DCX with 37 system calls (DCX has only 12 and DCE has 19). It runs in a round-robin time slice mode for equal-priority tasks, so tasks can be preempted by other equal-priority tasks when the time slice expires. In a sense, the operating system is a scheduler with interrupt features. Many limitations of DCX are gone: there can be 255 tasks with 255 different priorities, a whole host of mailboxes and semaphores as well as resource control. The operating system is basically a block of C language source, which you compile and link in with your own application code. Note that you supply the C compiler and write your tasks in either C or perhaps assembly. Versions are available for other processors and, like other products ported over from different processors, the fit with the 8051 is awkward.

USX is highly configurable: it is possible to omit the timing features and get a task sequencer, to omit event and resource controls, or even omit message capability (no mailboxes). The size of the various tables can be limited by restricting the number of tasks. These configuration choices reduce the size of the system code as well as the RAM usage. The code size is said to be about 10K for a fairly full system. It is not clear how the context switching is managed for the 8051. It is possible that the *entire* on-chip RAM is copied to off-chip space whenever tasks are switched. That would certainly leave a clear place for the next task, but would not be very fast. That could explain why the system is oriented toward large tasks with direct user-written interrupt handlers for quick actions. As long as an interrupt can avoid such a lengthy context switch, you save a lot of time.

CMX

The CMX Company markets a multitasking operating system which includes a version for the 8051. Like USX, it is a more general software product written in C and ported to the 8051. It has 47 system calls and is configurable much like USX. Context switching for the 8051 is variable depending on the particular C compiler used. You may want to check if the particular C compiler you use is supported since the wide variation in compiler tricks makes it difficult to have a generic operating system while handling nonreentrant functions.

In addition to message, resource, and event functions, CMX has functions to manage flags (bits within a single byte) and unique ones to manage lists. A distinction is made between a task and an interrupt; presumably the latter can be written by the user to quickly modify the appropriate event flags that signal the tasks. Unlike DCX, events are strictly task-defined and task-managed. An interrupt-related event must be set by an interrupt handler. Probably that would be done without context switching if possible. Both CMX and USX make a point of specifying a special procedure for calling a system function *from* an interrupt routine.

BYTE-BOS

Byte-BOS Integrated Systems offers an operating system that seems to straddle the large-small gap. It comes in versions specific to the particular C compiler being used (supports at least Franklin and Whitesmith). Like the previous two RTOS, it is primarily a C program, but it has part of the kernel coded in assembly for the 8051. It too is one of a family of RTOS for various processors.

Byte-BOS combines nonpreemptive and preemptive scheduling. The former allows tasks to run as long as they desire and then schedules the highest priority "ready" task when the running task puts itself to sleep. The task queue is a linked-list so there is no inherent limit on the number of tasks. The preemptive scheduler is invoked from an interrupt service routine (ISR). The interrupt routine overrides the currently running task and can "tell" the scheduler to suspend the interrupted task and execute higher priority tasks. It appears that Byte-BOS can emulate the simpler preemptive RTOS by adding a user-written ISR to cause rescheduling on every timer0 interrupt. It is stated that any function that places a task in the ready state will force rescheduling, so the preemptive nature of most system calls is present.

The documentation lists 42 system calls including timer, event (signal), message, resource, and interrupt management. Its complexity seems on a par with the other two big systems and the combination of round-robin and preemptive scheduling is fairly complicated at the start. Byte-BOS is unique in having an entire set of calls related to managing of serial communication via a UART.

RTXC

A. T. Barrett & Associates offer a real time kernel they call RTXC. It has been ported only to Franklin C for the 8051 so far, making use of the reentrant mode. Aside from possibly a small piece of assembly to handle the stacking of the registers, the operating system is written entirely in C and relies on the C compiler to produce the off-chip stack that is required. It does not appear that the on-chip RAM has been the object of any special attention for context switching, but the system will suffer from the same overhead of stacking for each system call as do several of the other systems.

Like most of the other systems, tasks of equal priority are scheduled round-robin while different priorities are preemptive. Task priority can be dynamically altered during operation. Without specific knowledge for the 8051 version, it is a reasonable guess that 254 different tasks are possible with a similar number of priorities—much is done in linked-list fashion. Obviously other limitations will get in the way of having a practical system that large!

The system is marketed in three versions and prices. The *basic* system has 24 calls but restricts the user to waiting for a single event at a time with no timeout on the wait. The *advanced* and the *extended* versions lead up to 56 system calls with ability to wait for multiple events at the same time. Interrupt routines can do nothing with the operating system except set semaphores for later processing.

While it has messages, queues, and partitions, RTXC has only a simple semaphore rather than a counting one. It is more like an event flag with three states: (1) *pending*—the associated event has yet to occur, or (2) *waiting*—the event has yet to come but a task is already waiting for it, or (3) *done*—the event has occurred.

The most unusual feature of the RTXC package is a system generation utility. All the configurable systems require the user to define the maximum number of tasks, mailboxes, semaphores, buffers, and so on, but RTXC provides a utility which prompts the user for all the necessary information and then produces .c and .h files. This doesn't relieve you of knowing the answers to all the questions, but it does reduce the chances for typing errors. You will want to be very clear about the names you give everything since there can be individual names for semaphores, mailboxes, memory buffers, queues, and tasks. You may need a scorecard for a large system!

OBSERVATIONS ON USX, CMX, BYTE-BOS, AND RTXC

These four operating systems are significantly larger and more flexible than the first three. They are also not designed for the 8051 alone and would allow your truly ANSI standard C code to be ported over to several different processor families. Although that may make them more general tools, it *seems* that they are not as carefully fitted to the 8051's unusual architecture and are less efficient (although I'm sure that would be argued by the various OS developers). All four have a significantly wider variety of system calls. The application programmer can choose to set up events, resources, memory pools, and so on. But, he also has to provide the interface to the specific hardware (including the standard 8051 features relating to interrupts, priorities, and timers). These may be provided in sample files or configuration examples, but the fact remains that there are more details to handle before getting started. The examples in the following chapters are only a comparison of the *working* part of the tasks; be aware that there may be a number of other configuration details to resolve before they will work.

REVIEW AND BEYOND

1. Define "real time." Describe two or three situations where "fast enough" is not very fast.
2. Define "multitasking," and compare it to parallel processing and conventional programming.
3. What are the advantages and disadvantages of multitasking over single-program solutions?
4. Why is the idea of a task helpful in project planning?
5. What is context switching? What feature was designed into the 8051 to speed context switching?

Chapter 9

Timing and Scheduling

KEEPING TRACK OF TIME

Many real-time applications depend on knowing the time. For example, the envelope handler of Chapter 1 needs to operate solenoids at the right time to synchronize the conveyors, the envelope stuffer, the sheet folders, and so on. Likewise, a system may need to measure the time between two inputs to determine motor speed. In real-time multitasking systems, time is usually overseen by a timer, which counts the system (crystal-stable) clock. This is entirely different from timing by software loops, although both ultimately rely on the system clock. Other things are going on while the timer counts; a software timer would not allow anything else to happen.

In its simplest form, timing can come directly from a real-time interrupt. In the "write-your-own" versions of the two examples in this chapter, timer 0 overflows every 10mSec, causing a regular interrupt. For short tasks, that is the only signal needed to produce a scheduler. By counting the interrupts you can determine at which "tick" each scheduled task should be called. As long as the task is done within the time before the next tick, there is no need for context switching (see Chapter 8) from the interrupt. Of course, any background task would need a separate context, but, assuming the two tasks are located to avoid variable overlap, two separate register banks would be sufficient.

Before going into the details, an outline of the chapter's examples will be helpful. There are only two examples; first they are both presented in C and PL/M *without* a multitasking **RTOS** (Real-Time Operating System). Then they are repeated in C using the operating systems introduced in the previous chapter. (Any of the stand-alone examples could be written in assembly as well, and DCX was originally interfaced only to PL/M [the C interface to DCX is a later addition using some assembly language glue]). If you plan to write *without* a commercial operating system, the first half of the chapters should suffice; if you are comparing various RTOS, study the second half closely; finally, if you have already made a specific RTOS choice, the chapter can be useful in getting started if the RTOS manuals are weak on examples. A *detailed* study of individual operating systems with extensive examples would require another book!

CAUTION: Aside from DCX, the *RTOS* examples are untested. What you see are best-effort solutions derived from the vendor's user's manuals for the respective systems with only brief review, if any, by the respective vendors. The intent is to illustrate the *similarity* of the systems for solving simple problems. It is recognized that significant features may not even be noted in the examples that follow. Perhaps a detailed treatment can be given when I have a better background of hands-on experience.

EXAMPLE: SOLENOID CYCLER

For a first example consider a scheduler, which turns on and off two solenoids attached to bits 0 and 1 of P1 in a fixed time pattern. Designate the solenoids as "S1" and "S2." At the cycle start, they are both off. Two seconds later S1 comes on. A tenth second later S2 comes on. S1 stays on for 2.0 Sec and S2 stays on for 2.4 Sec. The total cycle lasts for 4.5 Sec. A dynamically assigned cycle would be more realistic with user inputs for setup (like the envelope controller of Chapter 1), but this is simpler for a start. The feature that distinguishes the various versions of this example from conventional programming is the fact that it is keyed to a real-time interrupt. The intervals are determined by the match with the counter incremented in the real-time interrupt. The basic program flow is shown in Fig. 9–1a.

Figure 9–1b is a schematic for a solenoid driver circuit using an opto-isolated driver and a separate triac to drive the actual coil. Industrial solenoids often have 110vAC coils, so the interface is best done with optical isolation. If something failed at the output, you don't want the high voltage to destroy the entire controller. Also, the opto-isolation makes improved noise immunity more feasible. Figure 9–2 shows the actual code for the non-RTOS version.

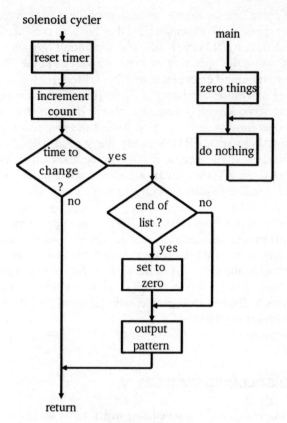

Figure 9–1a　Solenoid Cycler

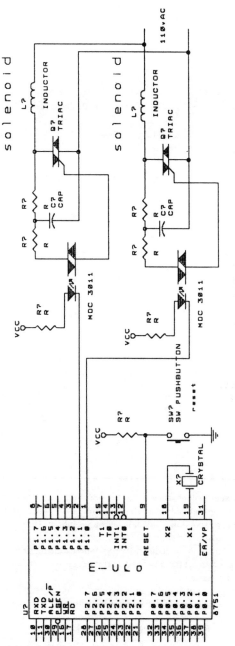

Figure 9–1b Solenoid Cycler

```
#include <reg51.h>
#define uchar unsigned char
#define uint unsigned int

uchar i;
uint nowtime;
struct code {uint abstime;uchar pattern;}next []=
    {{0,0x00},{200,0x01},{210,0x03},{400,0x02},{450,0xff}};

void cycletimer (void) interrupt 1 using 1{
    /*realtime interrupt routine
    happens every 10 mSec */
    TH0 = -8333/256; /*10mSec with 10MHz crystal*/
    TL0 = -8333%256;
    ET0 = 1;
    nowtime++;
    if (nowtime==next[i].abstime){
        if (next[i].pattern!=0xff) i=nowtime=0;
        P1=next[i++].pattern;
    }

}

void main(void){
    nowtime=i=0;
        TMOD=1;
        TH0=-8333/256;
        TL0=-8333%256;
        TR0=1;
        ET0=1;
    for(;;);   /*endless do-nothing background task*/
}
```

(Labels in left margin: 1 at struct code line; 2 at void cycletimer line; 3 at nowtime++; 4 at if (next[i].pattern line)

[1]This structure is a fixed-content schedule for the scheduler. In the next chapter, an expanded version is presented with a user-entered schedule. The first part of each structure element is the absolute time at which the corresponding solenoid states are sent out. Here only two solenoids are used, but the entire byte is available if desired.

[2]This interrupt is the key to the scheduler. Every 10 mSec the interrupt causes this interrupt routine to run. The first step is to reset the timer and start it out again so the minimum of time is lost before the next time interval is started.

[3]This variable is the counter of 10mSec units into the cycle.

[4]The end of the cycle is indicated by a 0 x f f for the pattern.

Figure 9–2a   Solenoid Cycler—C

```
SOLENOID:DO;
$INCLUDE (REG51.DCL)
    DECLARE FOREVER LITERALLY 'WHILE 1';
1   DECLARE NEXT (5) STRUCTURE (ABSTIME WORD,PATTERN BYTE)
    CONSTANT(0,00H,200,01H,210,03H,400,02H,450,0FFH);
    DECLARE I BYTE AUX;
    DECLARE NOWTIME WORD;

2   CYCLETIMER:PROCEDURE INTERRUPT 1 USING 1;
        /* REALTIME INTERRUPT ROUTINE EVERY 10 mSec*/
    TH0 =HIGH(-8333);/*10mSec(10MHz OSC)*/
    TL0 =LOW(-8333);
    ET0 = 1;
3   NOWTIME=NOWTIME+1;
    IF NOWTIME=NEXT(I).ABSTIME THEN DO;
4       IF (NEXT(I).PATTERN=0FFH) THEN DO;
          I,NOWTIME=0;
        END;
        P1=NEXT(I).PATTERN;
        I=I+1;
    END;
    END CYCLETIMER;

    I,NOWTIME=0;
    TMOD=1;
    TH0=HIGH (-8333);
    TL0=LOW (-8333);
    TR0=1;
    ET0=1;
    DO FOREVER; /*ENDLESS NULL LOOP*/
    END;
    END SOLENOID;
```

[1]This structure is a fixed-content schedule for the scheduler. In the next chapter, an expanded version is presented with a user-entered schedule. The first part of each structure element is the absolute time at which the corresponding solenoid states are sent out. Here only two solenoids are used, but the entire byte is available if desired.

[2]This interrupt is the key to the scheduler. Every 10 mSec the interrupt causes this interrupt routine to run. The first step is to reset the timer and start it out again so the minimum of time is lost before the next time interval is started.

[3]This variable is the counter of 10mSec units into the cycle.

[4]The end of the cycle is indicated by a 0x f f for the pattern.

Figure 9–2b   Solenoid Cycler—PL/M

EXAMPLE: PULSE GENERATOR

The example of Fig. 9–3 brings about a regular variable pulse width signal, which can be used to drive a motor or as a "poor man's" D-A converter. Depending on the repetition rate and the DC motor inductance, it is possible to vary the speed of the motor by varying the duty cycle of the drive. Such a signal can control speed with very little hardware—sometimes one transistor is enough—and the use of only one port pin. Incidentally, some of the 8051 relatives described in Chapter 7 have additional "PWM" (Pulse-Width Modulated) timers, which can provide such signals without regular software involvement. For slowly changing D-A outputs, the mere addition of a resistor and capacitor can deliver an analog voltage output by integrating the variable-width wave form. You will have to trade ripple against maximum rate of change, but many applications have no need for the fast performance of a normal D-A. Figure 9–4 shows the non-RTOS versions of the pulse generator example.

Incidentally, on many of the higher-performance "relatives" of the 8051, there are pulse width modulated (PWM) outputs which will run settable duty cycle and repetition rates without the continual involvement of the operating system and

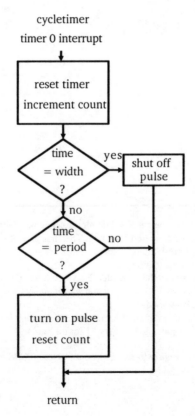

Figure 9–3 Pulse Task

```
   #include <reg51.h>
   #define uchar unsigned char
   #define uint unsigned int
   uchar i; uint nowtime;
1  uchar width=50; /*20% duty cycle*/
   uchar freq=250;

   void cycletimer (void) interrupt 1 using 1{
       /*realtime interrupt routine
        happens every 10 mSec*/
     TH0 = -8333/256;/*10mSec-10MHz crystal*/
     TL0 = -8333%256;
     ET0 = 1;
2    if (++nowtime==width) P1=0;
     else if (nowtime==freq){
       nowtime=0;
       P1=1;
     }
   }

   void main(void){
     forever{
     }
   }
```

[1]These variables are set to initial values so the program will do something on start-up. In Chapter 11, a similar version of the pulse function is used with a varying input from a control function.

[2]Nowtime is incremented *before* the test. If the time reaches the number in width the high part of the pulse is ended, while, if the time count reaches freq, a new pulse is started.

Figure 9–4a   Pulse Width—C

software. PWM counters are counters that derive their on and off times from different registers and may also have a prescaler to set the overall frequency. They can go much faster and with higher resolution than strictly software driven PWM systems.

While on the subject of analog/digital interface, it was mentioned in Chapter 1 that a V-F (Voltage to Frequency) converter can provide an excellent high-resolution A-D when speed is not important. In a similar way, a comparator can be added to a D-A to get an A-D—keep sending out different voltages until the comparator "flips." Then you know the digital value of the A-D into one side of the comparator is close to the code for the unknown analog voltage into the other side of the comparator. You can try values in an increasing count (a ramp) or in a "narrowing in" manner starting with the most significant bit (successive approximation). Sometimes, where you know the value doesn't change quickly, you can

```
PULSE:DO;
$INCLUDE (REG51.DCL)
  DECLARE FOREVER LITERALLY 'WHILE 1';
  DECLARE NOWTIME WORD, (FREQ,WIDTH)BYTE;

  CYCLETIMER:PROCEDURE INTERRUPT 1 USING 1;
       /* REALTIME INTERRUPT EVERY 10 mSec*/
     TH0 =HIGH(-8333);/*10mSec(10MHz OSC)*/
     TL0 =LOW(-8333);
     ET0 = 1;
     NOWTIME=NOWTIME+1;
     IF NOWTIME=WIDTH THEN P1=0; /*END PULSE WIDTH*/
1    ELSE IF NOWTIME=FREQ THEN DO;
       NOWTIME=0;
       P1=1; /*START NEXT PULSE*/
     END;
   END CYCLETIMER;
   I,NOWTIME=0;
2  FREQ=250; WIDTH=20;
   DO FOREVER; /*ENDLESS NULL LOOP*/
   END;
 END PULSE;
```

[1]If the time reaches the number in width the high part of the pulse is ended, while, if the time count reaches freq, a new pulse is started.

[2]These variables are set to initial values so the program will do something on start-up. In Chapter 11, a similar version of the pulse function is used with a varying input from a control function.

Figure 9–4b   Pulse Width—PL/M

start from the value last time and go up or down slightly depending on the result from the comparator (called a tracking converter).

The non-RTOS approach is much like the solenoid scheduler. Although the values of w i d t h and f r e q u e n c y are *fixed* in this example, they would normally be changed from outside by some other task using shared variables, signaling, or messages (Chapter 10).

EXAMPLE: SOLENOID CYCLER *WITH* RTOS

For a taste of the changes that come with an operating system, the solenoid cycler is rewritten using each of the operating systems mentioned in Chapter 8. The RTOS versions, shown in Fig. 9–5, calculate the incremental time delays from the absolute times; the earlier versions "woke up" every 10mSec and compared the absolute time, n o w t i m e, with the array of absolute times within the cycle. The key feature is the replacing of the real-time interrupt with a system call.

solenoid

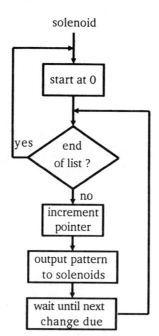

Figure 9–5a   Solenoid Task

With the DCX, DCE, and RTX calls, there are two parameters involved. One is the *timeout* count. If no other event comes within that time, the task will move to the *ready* state (if something else is using the processor) or *running* state (in the much more likely case that nothing else is going on) to toggle the next LED. The word "timeout" (actually a zero for DCX and DCE or an 8 with RTX) specifies that the task is to ignore any other possible events. Thus, by default, the wait will timeout. The time delays do not tie up the processor and the solenoids could go on operating seemingly quite independently of whatever else the processor might be doing.

In addition to the system calls, the use of DCX or DCE requires each task to have a block of identification information, which is called the Initial Task Descriptor (ITD). The ITD identifies, to the operating system, such things as where the task code (subroutine/procedure/function) begins, the amount of stack space to reserve, which interrupts to unmask, and what priority to assign to the task. It is a quirk of the languages that the ITD must usually be written in assembly code to get the linker to supply the correct address differences. The details should not be of concern here, but the DCX/DCE examples show the ITD at the top of the file—actually it is a separate module because it cannot be in-line assembled in PL/M or Franklin C.

The USX, CMX, Byte-BOS and RTXC examples are quite similar.

The one dramatic difference between operating systems is the size of system tick. DCX, DCE and RTXtiny all support 1mSec times, although DCE puts that at the lower limit. The other four carry enough overhead that they don't recommend ticks less than 10mSec. Perhaps by careful tailoring they can run faster, but the context switching is significant.

```
1  #include <dcx1.h>
   uchar i;

   void solenoid(void){
2      struct code {uint abstime;uchar pattern;}next []=
       {{0,0x00},{2000,0x01},{2100,0x03},{4000,0x02},
                                        {4500,0xff}};
       forever{
         for (i=0;next[i].pattern!=0xff;i++){
           P1=next[i].pattern;
3          rqwait(next[i+1].abstime-next[i].abstime,timeout);
         }
       }
   }


***********************************************************

   $INCLUDE(RMX51A.MAC)
   EXTRN CODE (solenoid)
   CSEG AT 0FFF0H
   ITD1:     5      6   7 8
4  %ITD(ENTRY1,10,81H,101B,2,0,ITD1+1)
   ENTRY1:LJMP solenoid
   END
```

[1]This include file (not shown) has the definitions of the system calls (like rqwait) as well as shorthand defines for uchar, uint, and forever. The custom-written interface between the C compiler and the DCX firmware is prototyped here.

[2]Values in an initialization statement may be missed where the C compiler does some initialization in a separate code piece that links with the main initialization. Where added tasks are linked in with existing firmware (iRCB 44/10 boards for example), the initialization code may not be reached where there is no main function. As an automatic variable, the array is filled by instructions in the function rather than a separate initialization.

[3]rqwait is the system call. Embedding the subtraction may obscure the fact that two parameters are passed. The first parameter is the timeout delay value in "system ticks" (normally 1mSec units with DCX). DCX allows only byte values for timeout for efficiency and speed. The system call as shown here in C actually uses an assembly module that interfaces the C code to the DCX firmware. Otherwise parameters are "passed" by preloading B and ACC—efficient but less "friendly."

[4]This is the initial task descriptor (ITD) which comes at the head of the task. It assembles to 11 bytes (plus the ljmp) and has to be done in assembly to work well with multiple tasks. The % is the ASM51 directive to include a macro—in this case it puts together a block of constants using the 7 parameters within the parentheses. The macro involved makes it easier to enter the parameters that make up the ITD. It is done in assembly because the PL/M compiler won't handle the subtraction of two addresses which are both unknown at compile time.

[5]The 10 is the amount of stack space reserved for this task (all the on-chip "real" stack).

[6]This task will use register bank 1.

[7]This task has a fixed priority of 2 (4=highest; 1=lowest). Unlike some larger systems, priority is not dynamically alterable.

[8]No external hardware or timer interrupts are used.

Figure 9–5b Solenoid Cycler—DCX & DCE

```
#include   <RTXtiny.h>
uchar i,status;

1  void solenoid(void) _task_ 0 {
       struct code {uint abstime;uchar pattern;}next []=
       {{0,0x00},{2000,0x01},{2100,0x03},{4000,0x02},
                                         {4500,0xff}};
       for(;;){
          for (i=0;next[i].pattern!=0xff;i++){
             P1=next[i].pattern;
2            status=os_wait(K_TMO,
                      next[i+1].abstime-next[i].abstime);
          }
       }
   }
```

[1]RTXtiny initializes the task numbered zero automatically. Other tasks need to be created in task 0.

[2]os_wait is the system call. Embedding the subtraction obscures the fact that two parameters are passed. The first parameter (a byte) is the mask which identifies the possible event(s) to awaken the task. The second parameter is the delay value (here system ticks assumed to come at a 10mSec rate).

Figure 9–5c   Solenoid Cycler—RTXtiny

```
#include <usxcfg.h>
#include <usx.h>
uint status;
uchar xdata free_memory[MEM_ALLOCATION];
uchar solslot;
uchar i;

void solenoid(void){
   struct code {uint abstime;uchar pattern;}next []=
   {{0,0x00},{200,0x01},{210,0x03},{400,0x02},{450,0xff}};
   for(;;){
      for (i=0;next[i].pattern!=0xff;i++){
         P1=next[i].pattern;
1        dlytsk(solslot,1,
                  next[i+1].abstime-next[i].abstime);
      }
   }
}

void main(void{
   MEM_DEF *memptr;
2  usxini();
   usrclk_init();
   memptr=(MEM_DEF *)free_memory;
   memptr->mem_size=MEM_ALLOCATION-sizeof(struct mem_def);
   memptr->task_slt=SYSTEM;
   memptr->mem_nxt=NULL;
   relmem((usxmem_t *)(++memptr));
3  solslot=runtsk(5,&pulse,20);
4  begusx();
}
```

[1]The delay has three parameters. The first is the slot of the particular task to be delayed (here the given task, arbitrarily in slot 5). The second parameter is the type of delay—the 1 indicates system ticks while 2 and 3 indicate units of seconds or minutes, respectively. The third term is the number of units of delay—here the number of ticks until the next solenoid state change. It is assumed that the clock is initialized for 10mSec units.

[2]All these details are involved with initialization of the operating system—where it can keep its variables, tables, and so on.

[3]This actually puts the solenoid task on record with the RTOS. The 5 is the priority with zero lowest and 255 the highest priority. The second parameter is a pointer to the task. The final parameter is the stack size—here kept fairly small. The returned parameter is the slot number assigned to the task.

[4]This actually starts the operating system. There should never be a return.

Figure 9–5d   Solenoid Cycler—USX

```
   #include <cmxfunc.h>
   uchar solenoidslot;

   void solenoid(void){
      struct code {uint abstime;uchar pattern;}next []=
       {{0,0x00},{200,0x01},{210,0x03},{400,0x02},{450,0xff}};
      uchar i;
      for(;;){
         for (i=0;next[i].pattern!=0xff;i++){
            P1=next[i].pattern;
1           cxtwatm(next[i+1].abstime-next[i].abstime);
         }
      }
   }

   void main(void){
      cmx_init();
2     cxtcre(4,&solenoidslot,0,solenoid,20);
3     cxttrig(solenoidslot);
4     cmx_go();
   }
```

[1]`cxtwatm` is the system call. It actually returns an unsigned status character, but since no testing is being done, it is omitted from the system call. It would indicate either the timeout or else that some other task, timed procedure, or interrupt caused the task to resume—not the intention here. Embedding the subtraction to get the timeout may obscure the fact that only one parameter is passed—the timeout delay value in system ticks. The value is an unsigned integer.

[2]This creates the task of priority 4 (0 is highest, 255 is lowest) with zero message envelopes (it receives no messages) and having a 20-byte stack space. The stack size is not likely to need to be larger since no library functions are used.

[3]This puts the task in the ready state so it can run when the operating system starts.

[4]This starts the operating system. There is no return.

Figure 9–5e Solenoid Cycler—CMX

```
#include <bbdefs.h>
BB_TCB_TYPE solenoid_tcb;
uchar xdata xstack[256];
uchar xdata context[64];
uchar data istack[20];
uchar i;

void solenoid_task(void){
   struct code {uint abstime;uchar pattern;}next []=
   {{0,0x00},{200,0x01},{210,0x03},{400,0x02},{450,0xff}};
   for(;;){
      for (i=0;next[i].pattern!=0xff;i++){
         P1=next[i].pattern;
1        bb_delay(next[i+1].abstime-next[i].abstime);
      }
   }
}

void main(void{
2    bb_init_bos();
3    bb_cteate_task(&solenoid_tcb,context,
                          istack,xstack,0,0,1);
4    bb_run_task(&solenoid_task,solenoid);
     bb_start_bos();
}
```

[1]This call delays with no sensitivity to other events. It is equivalent to a simple timeout in some other systems. The event-sensitive call, wait, is used in the next example.

[2]This initializes the operating system.

[3]The create_task call requires all sorts of details from the user. Specifically you must define a task control block (tcb), a context storage block (array), both internal and external stack blocks, a message buffer and buffer length (here set to 0 since we use no messages), and the priority of the task (here zero, which is highest priority).

[4]This places the task in the ready state (but does not start until the operating system is started with the bb_start_bos call).

Figure 9–5f   Solenoid Cycler—Byte-BOS

```
#include<rtxcapi.h> /*rtxc KS prototypes*/
#include<cclock.h>
uchar i;

void solenoid_task (void) {
    struct code {unit abstime;uchar pattern;}next [] =
    {{0,0x00},{200,0x01},{210,0x03},{400,0x02},{450,0xff}};
    for(;;){
        for (i=0;next[i].pattern!=0xff;i++) {
            P1=next[i].pattern;
            KS_delay(0,next[i+1].abstime-next[i].abstime);
        }
    }
}
```

1

¹The K S_d e l a y is a simple timed delay ("blocking") of the task indicated. Here the zero indicates that it is the calling task which is to be delayed, but it is quite possible to block *other* tasks for a specified time.

Figure 9–5g   Solenoid Cycler—RTXC

EXAMPLE: PULSE TASK *WITH* RTOS

First of all, let it be pointed out that the pulse could quite easily be created by two timeouts in succession where the first was for the pulse width and the second was for the off time! A simple subtraction should handle it. But, in order to show the use of a *separate* regular interval event, this example is done in what is certainly a more complex way for some of the systems. No attempt is made to assess the impact of the different approaches in a more heavily loaded system where preemption could interrupt tasks and throw off timing. Specifically, does the pulse recover lost time later if the task can't run right away, or does the system slip all the following pulses?

The examples, shown in Figure 9–6, could be set to directly copy the non-RTOS versions, but to show the flexibility of the systems, the timeout and the periodic interval event are combined where possible to allow the pulse *frequency* to be maintained by one value (the interval value) while the pulse *duration* is maintained by another (the timeout value). An *interval* should be a high-priority event that comes at a regular, repeated time regardless of whether the task using the interval is running or not. A *timeout*, on the other hand, does not begin counting until the task does a system call.

The periodic timing is handled quite differently in the various systems. Software for a periodic *event* is built-in for DCX and DCE (s e t i n t e r v a l), USX (p e r i o d), Byte-BOS (e n a b l e_s a m p l i n g) and RTXC (s t a r t_ t i m e r). RTX and CMX are quite different. For RTX, the user writes a *separate*

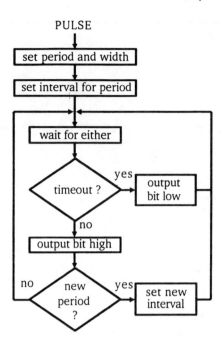

Figure 9-6a   Pulse Task

task that produces a signal for the task in question. For CMX the user writes a timed procedure (not a task which is an endless loop) which starts and runs to completion each time the time runs out (or just once, if you set a parameter differently). This requires more user code, but may be no problem where many tasks are allowed to exist at the same time. This is a somewhat like Byte-BOS with high- and low-level system calls—you can build up a periodic event from another task or procedure.

RTXC seems a little more complicated because it maintains a pool of timers which may become exhausted. Timers are allocated and freed dynamically, so it is possible to have a situation where no timers are available. In general it is recommended that a task allocate all its needed timers before entering the forever loop. This will guarantee the timers will be available when needed. Unlike the other operating systems, *several* timers can be running at the same time for one task.

Incidentally, CMX requires you to write the system tick also. It is included only in the first example of Chapter 11 where interrupts are discussed.

```
$INCLUDE(RMX51A.MAC)
EXTRN CODE (solenoid)
CSEG AT 0FFF0H
ITD1:
%ITD(ENTRY1,10,81H,101B,2,0,ITD1+1)
ENTRY1:LJMP pulse
END
```

```
****************************************************

  #include <dcx1.h>
  void pulse(void){
  uchar width=50; /*20% duty cycle*/
  uchar freq=250;
  uchar oldfreq=250; /*4/second*/
  rqsetinterval(freq);
  forever{
1     rqwait(width,timeout|interval);
2     P1=(status)?0:1;
      if (freq!=oldfreq){ /*changed*/
        rqsetinterval(freq);
        oldfreq=freq;
      }
    }
  }
```

[1]This is the wait for either a timeout or an interval. The interval comes every 250 mSec and signals the start of the next pulse. The timeout (f r e q) comes when the pulse is done. The timeout may come several times before the next pulse starts, but zero is still zero no matter how many times it is sent.

[2]This line sets the port high if status is greater than zero (here an interval) or low if status is zero (a timeout).

Figure 9–6b Pulse Task—DCX & DCE

```
   #include <RTXtiny.h>
   uchar i,status;
   uchar freq,width;

1  void interval(void) _task_ 0{
       freq=250;
       width=50;
2      os_create(1);
       for(;;);
3          os_wait(K_TMO,freq);
4          os_send_signal(1);
       }
   }


5  void pulse(void) _task_ 1 {
       for(;;){
6          status=os_wait(K_TMO | K_SIG,width);
7          P1=(status==K_TMO)?1:0;
       }
   }
```

[1]Task 0 starts up automatically on start of the operating system (which is all handled by the material in the header file). There is no m a i n function to write. This task makes the interval which is used as a signal to the pulse task.

[2]The first task to start must start up at least one of the other tasks.

[3]This is a simple wait for a timeout at f r e q system ticks. It is used to make an interval for the pulse task because no resources exist to do it in a system call.

[4]This causes a signal (event) for task 1 (the pulse task). The signal is discussed further in the next chapter, but the actual effect is the setting of a bit in the signal byte for task 1.

[5]The pulse task waits actually drives the lsb of P1 high for "width" system ticks and low for the rest of the "freq" ticks that make up the repetition period. Technically, f r e q should be called something like "period" since it is the inverse of the frequency.

[6]The status byte which is returned is used here to determine whether the task timed out or received the interval signal from the other task.

[7]The ( ) ? : operator condenses the i f ( ) ; e l s e ; operations. If the test is true then it timed out and the port is set low because the width time has expired. Otherwise it is set high to start a new pulse.

Figure 9–6c   Pulse Task—RTXtiny

```
  #include <usxcfg.h>
  #include <usx.h>
  #define period_event 1
  uchar width=5;  /*20% duty cycle*/
  uchar freq=25;
  uchar oldfreq=25;  /*4/second*/
  uchar pulse_slot; uint status;
  uchar xdata free_memory[MEM_ALLOCATION];

  void pulse(void){
1     period(period_event,freq);
      for(;;){
2         status=wteset(period_event,width);
          P1=(status==period_event)?1:0;
3         if (status==period_event) decevt(periodevnt);
          if (freq!=oldfreq){ /*changed*/
              period(period_event,freq);
              oldfreq=freq;
          }
      }
  }

  void main(void{

      /*same initialization as Fig.9-5c */

      pulse_slot=runtsk(5,&pulse,20);
      begusx();
  }
```

[1]This system function sets an event, here arbitrarily chosen to be event #1 (described by a define to make more readable code) to be set every 10 system ticks. The assumption is that the USX system is configured with a system tick time of 10mSec.

[2]At this point the task will wait for either a timeout number of system ticks (width) or for the period event (the period of the wave form being generated by the task).

[3]If a period event caused the task to resume, in USX it is necessary to clear the event. Actually, per i od_event is a one-byte *counter* which is incremented every "freq" system tick and this line of code subtracts one from the count to cancel the increment done by the period function. By counting rather than using a single bit flag, it is possible to get several events behind in processing and still not lose any of them.

Figure 9–6d Pulse Task—USX

```
1  #include  <cmxfunc.h>
2  uchar  slot,events,timer;
   uchar  width=5;  /*20% duty cycle*/
   uchar  freq=25;
   uchar  oldfreq=25;  /*4/second*/

3  void interval(void){
      status=cxesig(events,1,1);  /*interval event*/
   }

   void pulse(void){
     for(;;){
4        status=cxewatm(events,1,1,1,width);
5        if (status){ /*interval event set*/
            P1=1;
6            status=cxesig(events,0,1);  /*clear interval event*/
         }
         else P1=0;  /*timeout=pulse width*/
         if (freq!=oldfreq){ /*change in interval needed*/
7            cxtpstt(timer,freq);
            oldfreq=freq;
         }
      }
   }

   void main(void){
     cmx_init();
```

[1]This header file prototypes all the system functions.

[2]These are the various variables used by the operating system.

[3]With CMX the interval causes a procedure to run. In order to get an *event* for the task, the user must write a procedure to produce the event, shown here. CMX behaves differently for periodic timed tasks. The event group is a byte which can hold up to 8 event bits. The task can be awakened by any combination of events, but all the events have to be user defined and must be managed by the user-written tasks. In this case, a periodic-procedure system call is used to make the regular interval event and only one event is used in the group. Notice that the periodic procedure is not in a "forever" loop; apparently the operating system picks up when it finishes.

It would be possible to produce the same result without the interval, perhaps, by simply using a simple wait for the pulse high time and a second wait for the difference between freq and width. This is used in the last example of Chapter 11.

[4]This is an event wait with a timeout of "width" time slices. The task will wait for the first bit of the event or will timeout in width ticks. The status will be the key to deciding which happened.

[5]A zero indicates a timeout so here anything else has to be the occurrence of an interval event.

[6]Once an event has occurred, it must be manually cleared, as shown here.

[7]The interval timer must be reset if the frequency value is changed. Here it would have to come from some other task changing the value. Ultimately it would be done better with a message.

Figure 9–6e   Pulse Task—CMX

```
8       cxtcre(4,&slot,0,pulse,20);
        cxttrig(slot);
9       cxtpcre(&timer,&interval,freq);
        cxtpstt(timer,freq);
        cmx_go();
    }
```

[8]This line creates the pulse task—puts it on record with the operating system.
[9]This pair of system calls first sets up the periodic interval timer within the system and then starts it. It even has control of how long to wait before the *first* time. The interval is the frequency for the wave form.

Figure 9–6e Continued

```
#include   <bbdefs.h>
#define interval 0x0001
BB_TCB_TYPE pulse_tcb;
BB_TCB_TYPE interval_tcb;
uchar xdata pxstack[256];
uchar xdata ixstack[256];
uchar xdata pcontext[64];
uchar xdata icontext[64];
uchar data pistack[20];
uchar data iistack[20];

uchar width=5; /*20% duty cycle*/
uchar freq=25;
uchar oldfreq=25; /*4/second*/

void interval_task(void){
1       bb_enable_sampling(freq);
        forever{
2         bb_sleep();
3         if(bb_timeout_occurred()){
            bb_ack_timeout();
4           bb_post_event(&pulse_tcb,interval);
```

[1]Contrary to the first impression, ByteBOS doesn't allow a task to have both a sampling interval and a timeout running at the same time (sampling is the term used for a periodic event to cause the task to wake up). The solution, as with RTX and CMX, is to make a *separate* task which manages the interval and then signals the pulse task. Getting the regular time could have been done with de l a y here, but it was decided to illustrate the variety of "low-level" calls in the operating system. The e n ab l e_s amp l i ng causes a periodic event for *this* task.

[2]The sleep here is the generic one which returns for any possible event. Several other versions exist where the task wakes up only if *specific* events occur.

[3]The status test has the effect the name implies. Since no other event is possible for this task, this may be unnecessary.

[4]This call will signal the pulse task that it can start a new pulse. The event number was defined with the word "interval" to give more clarity; in this case it was arbitrarily defined as event #1.

Figure 9–6f Pulse Task—Byte-BOS

```
5            if (freq!=oldfreq){ /*changed*/
               bb_sample(freq);
               oldfreq=freq;
             }
           }
         }
      }


  void pulse_task(void){
6      bb_enable_event(interval);
       forever{
7        status=bb_wait_for_ event(interval,width);
8        P1= (status)?1:0;
       }
     }


  void main(void{
       bb_init_bos();
       bb_cteate_task(&pulse_tcb,pcontext,
                          pistack,pxstack,0,0,1);
       bb_cteate_task(&interval_tcb,icontext,
                          iistack,ixstack,0,0,0);
       bb_run_task(pulse_task,&pulse_tcb);
       bb_run_task(interval_task,&interval_tcb);
       bb_start_bos();
     }
```

---

[5]Notice that f r e q would never change unless it was shared with some other task. The messages of the next chapter would be a better way to handle this; a message could come from an input task tied to a keypad.

[6]This task has to be set up to accept events. Here it is actually only allowing one event with the mask of 0x0001.

[7]All the other programming was to get this line to work! The interval is the event for which the task waits and the width value is the timeout delay. In this system it might be simpler to solve this problem by computing an off-delay by subtracting w i d t h from f r e q and using two simple d e l a y calls.

[8]The zero status indicates the task did time out and the on time of the pulse should end (a zero out port P1). Otherwise, since the periodic event called "sampling" is the only other event for this task, it is time to start a new pulse.

Figure 9–6f Continued

```
#include <rtxcapi.h>
#include <cclock.h>
#include <csema.h>
SEMA semalist{}={semperiod,0};
CLKBLK *periodtimer

uchar width=5; /*20% duty cycle*/
uchar freq=25;
uchar oldfreq=25; /*4/second*/

     void pulse_task(void){
1       KS_start_timer(periodtimer,freq,freq,semperiod);
        forever{
2         status=KS_waitt(semperiod,width);
3         if (status==RC_GOOD) P1=1;
          else if (status==RC_TIMEOUT) P1=0;
          if (freq!=oldfreq) { /*changed*/
4           KS_restart_timer(periodtimer,freq,freq);
            oldfreq=freq;
          }
        }
5    }
```

[1]The timer is allocated with a connection to the associated semaphore, semperiod. The dual entry of freq indicates that the event should come the first time after freq clock ticks and then repeat every freq ticks.

[2]The waitt is one of the "advanced" system calls which allows a wait for an event (here the timer started earlier) or a timeout.

[3]Like the other examples, the pulse is high at the beginning of each period and set low after each timeout has expired. To be completely safe, a test could be included for the discovery that there are no timers free to do the timeout—hopefully you would have configured the system with enough of them. If you wished to avoid the high-level system calls (which come only with a more expensive operating system package) you could do a simple KS_wait for the timer and then do a KS_delay for the pulse width.

[4]The restart is a high-level function that combines a KS_stop_timer with a KS_start_timer.

[5]The initialization is done in a configuration utility which asks about priorities, timers, and so on. Don't think the omission here relieves you of those system questions!

Figure 9–6g Pulse Task—RTXC

REVIEW AND BEYOND

1. What happens if one uses a time-burning delay loop under an RTOS? Why is it not a good idea?
2. What is the default basic unit of time for DCX? Are there typical values for other RTOS?
3. What is the difference between a timeout and an interval? When would an interval be preferable?
4. What are real life activities that correspond to timeout and interval timing?
5. What if LEDs were replaced with incandescent lamps? How might that affect the maximum flashing rate due to the time to achieve brilliance?

Chapter 10

Communication and Synchronization

TASK COMMUNICATION

The Chapter 9 examples involve separate, unrelated tasks, which begin at set times. To produce a flexible, adaptive system while still maintaining separate tasks, it is necessary to have communication *between* tasks. This chapter illustrates the use of shared variables and flags, the use of more formal signaling, and the use of message passing techniques of RTOS.

SHARED VARIABLES FOR COMMUNICATION

Chapter 6 on modular programming demonstrated sharing of variables between *modules*, but this takes on new significance in the context of tasks. Where each task is written as a separate module, it is easy to isolate the various functions of the system. By controlling the scope of variables in programming modules, you can have modules *intentionally* use the same name for the same thing. The linker will then assign the variable in each task to the same location. Using shared variables, *one* task might read an input from some switches indicating the speed a motor should turn, and leave the number (perhaps in a processed form representing the desired delay time between tachometer pulses) in a shared location named target. A *separate* (interrupt-driven) task could determine the actual time between pulses and leave an updated value in actual. A third task could compare target with actual and make an appropriate adjustment to drive, a variable used by a pulse width motor drive as in the second example of Chapter 9. In that way the input and drive-adjusting tasks could handle any processing details in a low-priority mode while the moment-by-moment driving of the motor and reading the tachometer could be done at a higher priority without the delay of processing and math routines.

There are some drawbacks to communicating with shared variables. First, the receiving task has no idea when a new piece of information has been sent. For example, the case above would not know directly when the tachometer value or the switch value has been changed. The computation might have to cycle endlessly regardless of whether or not a new piece of data has arrived. In multitasking terms, the updating of information is not an *event*. If the receiving task ought to promptly begin some activity when information arrives or changes, it must frequently check to see if there has been a change. That in itself takes up processor time. But a second, bigger problem in the multitasking perspective is that there is no protection from loss of information by having one task change it in the middle of some other task's use. If a low-priority task is updating a string of variables when a high-priority task comes along and modifies the string, the latter task could get part new and part old information. You could set a flag variable to warn the high-priority task to keep out until done, but you will soon have written the equivalent of your own operating system! Unfortunately, this characterizes the evolutionary development of most embedded applications. When it is finally realized that a commercial multitasking operating system would be better, there is already too much invested in software development to tolerate a change.

SEMAPHORES AND SIGNALING

A bit or a counter can be used to synchronize various tasks. Like the railroad semaphore, which is a signal to a following train to slow up or wait because a train has recently passed a stretch of track, so the bit or counter can indicate that a task should wait until some other task is done. For the railroad semaphore the *event* is the preceding train entering or leaving the signaling block. In multitasking the event can be the intentional setting or clearing of the bit or counter by one task that signals the other task to wake up. Chapter 9 already discussed time-related events—timeouts and intervals. For some examples it was actually necessary to write a separate task that would time the interval and signal the interval event to the other task. In those cases some intertask signaling is used, whereas the other systems have operating system features to produce the time events directly.

This chapter describes other signaling between tasks where the signal may have *any* mutually agreed upon significance. In its simplest form, signaling is by any shared bit variable. A signal could indicate that a task has finished processing a block of memory and an acquisition task is free to refill the block, or it may be a signal from X-Y table driver tasks that they are done and the drill-lowering task is free to drill the next hole in the printed circuit board. Whatever the signal, it must be consistently interpreted by the various tasks using it.

Signaling can involve more than a bit (specifically with USX), in which case the semaphore is a *counter*. Terms like "suspension depth" and "units" come into play where the more units in the semaphore the greater the suspension depth. With the printed circuit drilling system, for example, the start-up task could put *two* units into the semaphore and the X and Y tasks could each remove *one* unit when they have finished. The UP-DOWN task would have a suspension depth of two and would wait until there were zero units in the semaphore before running. In that way, you avoid a separate task to manage which tasks gets done first and when to proceed. As you may see, this is quite flexible, but it requires the imagination of the system planner to determine which operating system resources to use.

MESSAGES

Although it is possible to share entire arrays or structures between tasks just as bits and single variables are shared, most multitasking systems can exchanging information by passing **messages**. The analogy is made with letters arriving in a mailboxes. Messages, like letters, pile up until you get them out of the mailbox. Without yet going into the details of the message itself, consider the benefits of messages over shared variables:

1. Messages are *events*—a task can go to sleep *until a message arrives*. This allows tasks to start up only when the message arrives with no need to periodically use processor time to check the mailbox.
2. Tasks using messages do not have to be linked together. For DCX and DCE, communicating tasks can even be on different boards with no knowledge of each other, as will be discussed in Chapter 12. There is no need to know where the variables are in memory.
3. Messages (usually) queue up for a task, and a later message does not overwrite an earlier one. Messages can be received in succession by the task. Unlike shared variables, messages do not get overwritten.

COMPARING COMMUNICATION RESOURCES

There are some significant differences between message and signaling mechanisms in the various RTOS.

DCX and DCE both automatically have just one mailbox per task and the messages go directly to tasks. Except for off-board messages, the only information that moves is the pointer to the start of the message. A 7-byte header carries information about the destination (address) and the source (return address) of the message as well as the length. The sending task must consider the RAM space as unavailable once the message is sent; otherwise it could corrupt the data before the receiving task has used it. The messages on-board can be in on-chip or off-chip RAM and the former can be of essentially any length. The latter are critical to off-board communication and are typically set to a maximum of 20 bytes including the header.

DCX and DCE have no signaling mechanism as such, but empty messages can serve a signaling purpose.

RTXtiny has no message mechanism and would have to alert a task to a message by use of signaling discussed above.

Signaling with RTXtiny is also limited. Each task has only one signal bit. Each task wishing to send a signal may send the signal to one specific task at a time. Clearly this is quite simple and must be fast to execute, but the complex options of the following systems are not going to be easily reproduced.

RTX apparently has eight mailboxes total with fairly short messages, but the details were not available at the time of writing.

USX passes a pointer to the message, which can be any arbitrary array or structure. The message pointer is sent to a mailbox, of which there can be any number configured at configuration. There is no specific tie between a task and a mailbox. Since only the pointer to the message moves, the sending task can only know if the actual message data space is free for reuse by a return signal from the receiving task or by use of memory buffers, which can be requested and released. Essentially, like DCX and DCX, the user has to ensure the "envelope" is free for a new message. USX adds a unique feature where the messages to a mailbox are assigned priorities and the first-in-first-out aspect can be overridden to have a high-priority message picked up first by the task "checking the mail."

Events in USX can hold numbers besides 0 and 1—they are closer to counters of events. It is not clear whether they count to 8 bits or 16 bits, but they are created at configuration and belong to no particular task. Any task(s) can set, clear, decrement, or increment them and can wait for a zero or nonzero condition to come to any particular flag. Event *groups* are a different thing—a 16-bit set of flags which are either set or cleared. You can mask to wait for any particular combination of specific bits before proceeding with the given task. As with the next two systems, the options become rather mind-boggling!

CMX is the only system actually copying the *contents* of the message when it is sent. The default length (envelope size) is 12 bytes, but it can be as small as 1 byte if so initialized. For efficiency the system can be configured for 2 or 3 bytes which serve as pointers to the actual messages. All envelopes are configured the same size. There are several variations of sending and receiving depending on whether the sender should go to sleep, wait for a reply, wake up a sleeping destination task (have the mailman ring the doorbell?), and so on. Likewise, the receiving task can sit and wait until a message arrives, wait for a specified time, or just check the mailbox with delay if no message has arrived. The large variety of choices allows the system to perform actions that would otherwise require several different calls.

CMX has two categories of events that need to be distinguished. Flags relate to a specific task while events are more separated signals that can relate to all the tasks. Both flags and event groups are 8 bits wide. Each task automatically has a single flag byte while any number of event groups must be created. The waits for flags or events involve a mask and a condition test so you can wait for only specific bits being high or low while ignoring all the rest. There are a number of system calls because there are waits with and without timeouts both for flags and events. There are also calls that can wake another task which has suspended itself—a directed signal with no conditions. The options are so many that you will probably go through life using only some of the commands without noticing the lack of the others. There are differences between calls which, much like the choice of variable type, can affect efficiency, but the job can be done several ways. Very careful planning is suggested before using event groups with multiple tasks.

Byte-BOS has what it calls "low-level" and "high-level" functions. The distinction is that low-level calls do not suspend the task—either the message has already arrived or the call returns with the indication that no message was waiting. High-level calls can wait for the message to come (or timeout). The high-level call `sleep_for_message` is a `wait_for_message` without provision for a timeout.

In addition to messages, the Byte-BOS has provision for user-defined events (signaling). There are at least 16 event bits, and events can be selectively posted to one task or all tasks can be notified of the same event. Each task automatically gets just one event "holder" (integer). Likewise, one task can wait for *any* of a number of event bits to be set or it can wait for *all* of a specific set of event bits. It will take some time to really use the options to their fullest, but the number of calls will seem larger than with some other systems because the various options within the same function are broken out as separate calls.

RTXC can have both fixed location and allocated messages in the sense that a message could be transferred to a memory block before being sent or it could be sent directly. Since nothing except a pointer moves, the sending task should not reuse a message block until it is sure the receiving task has obtained the information. With allocated blocks, on the other hand, it is possible for the sending task to obtain another buffer and begin preparing another message at once without concern about synchronization with the receiving task. There are asynchronous (don't wait), synchronous (wait for receipt of the message and an acknowledgement), and conditional (wait for a reply with a timeout) messages. On the receive end there are also three options: (1) polled receipt which goes on if no messages are in the mailbox, (2) unconditional receipt which waits forever for a message, and (3) conditional receipt which gives up waiting after a timeout period. Messages can have priorities so urgent ones can come out of the mailbox before ones arriving earlier. With synchronous message exchange it is possible for the receiving task to retrieve the information, change the contents of the message, and return it as a reply for a convenient two-way message exchange. Like all the systems, the arrangement and significance of the message contents is determined by mutual agreement between the tasks.

RESOURCES, POOLS, AND LISTS

In some of the operating systems there are options for management of resources, lists, queues, and memory pools. These features differ between systems and are essentially somewhat more automatic signaling managed by the operating system to keep one task from accessing variables (or ports or other hardware) before

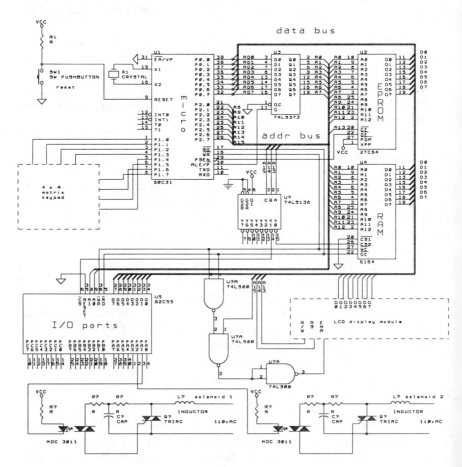

Figure 10–1a Solenoid Sequencer II

another task is done. They may do the specific function more efficiently, but the equivalent function can be built up from the other system calls and additional programming. Specifically, DCX and DCE have a memory pool manager of *on*-chip RAM; DCE, USX, CMX and RTXC have a memory pool manager for *off*-chip RAM management; Byte-BOS has a unique UART buffering system; CMX and RTXC have a list/queue manager; and USX, CMX, Byte-BOS and RTXC all have a resource manager. Incidentally, where possible the example that follows illustrates resource management in accessing the LCD display.

EXAMPLE: SOLENOID SEQUENCER II (KEY ENTRY)

By revising and adding to the solenoid sequencer of Chapter 9, including an LCD display task, adding a control task, and revising the keyscan task in Chapter 2, a key-entry sequencer programmer is produced. Figures 10–1a through 10–1b show the hardware and the interrelations of the program pieces. The task block diagram of Fig. 10–1b will be particularly helpful in visualizing the interaction. Flow charts for the tasks are in Fig. 10–2a. Figures 10–2b and 10–2c show the non-RTOS software in C and PL/M.

Part of the main function or another task, i n i t i a l i z e, sets up the LCD display module and registers. The details are of no particular concern here—it is just a succession of writes to the LCD module, which happens to be a standard 2 × 40 module.

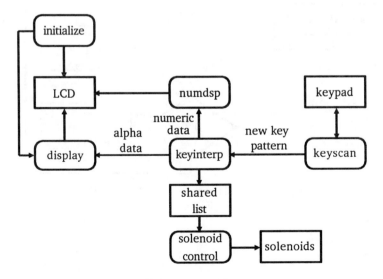

Figure 10–1b   Solenoid Sequencer II

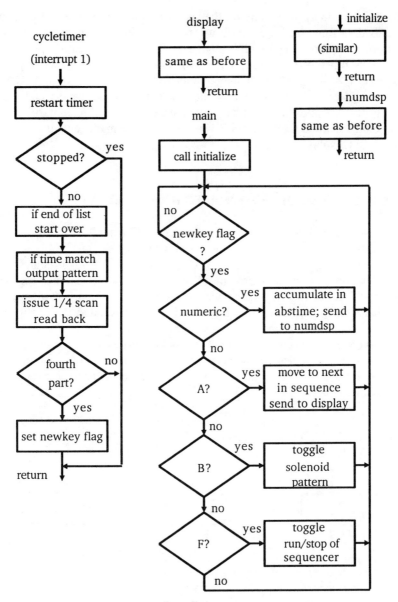

Figure 10–2a   Solenoid Sequencer II—*non* RTOS

```c
#include <reg51.h>
#define forever for(;;)
#define uchar unsigned char
#define uint unsigned
#define lcddat XBYTE[0xc000]
#define lcdcmd XBYTE[0x8000]
#define pcmd XBYTE[0x2003]
#define PORTA XBYTE[0x2000]
```
1
```c
bit newkeyflg,newnumbflg,goflg;
```
2
```c
struct {uint abstime;uchar pattern;} next [20];
uchar i,j,entry,x;
uint nowtime,scan,oldsc,newsc;
```
3
```c
void display(uchar startloc,uchar *s){
    lcdcmd = 0x80 | startloc; msec(1);
    while(*s){
        lcddat = *s++; msec(2);
    }
}
```
4
```c
void numdsp(uchar startloc,uint number,bit dpt){
    lcdcmd = 0x80 | startloc; msec(1);
    uint m = 10000;
    bit ldgzero = 0;
    do{
        if ((((number/m)>0) || (m == 1))
            || ((m == 100) && (dpt == 1)))
        ldgzero = 1;
        if (ldgzero == 1)
        lcddat = number/m+'0';
        else lcddat = ' ';
        msec(1);
        number %= m;
        if ((dpt == 1) && (m == 100)) {
        lcddat = '.';
        msec(1);
        }
    }
    while((m /= 10)>0);
}
```

[1]The signaling between functions is handled largely by flags which are public shared variables. Particularly n e w k e y f l g lets the interpretation function know it can proceed to process a new input without tying up the interrupt function for a possibly long time.

[2]This structure is more flexible than that of Chapter 9. It could be made *based* and put in memory space above that claimed by the located program. Such storage can even be managed for you by some operating systems. These functions would be the same as in the first example and are prototyped as a reminder that they would be included by linking.

[3]This function is simpler because, by default, a generic (3-byte) pointer is used and the third byte notifies the function which space is to be used.

[4]This function may be useful to borrow, but it is too complex to absorb quickly. It takes an unsigned integer and converts it to a decimal display on the LCD. It suppresses leading zeros but maintains alignment by substituting blank spaces.

Figure 10–2b   Solenoid Sequencer IIC

```
5  void cycletimer(void) interrupt 1 using 1{
        /* realtime interrupt routine every 10 mSec*/
        uchar code scanpat[]={0x10,0x20,0x40,0x80};
        TH0 =-8333>>8;/*10mSec(10MHz osc)*/
        TL0 =-8333%256;
        ET0 = 1;
         /*solenoid drive portion*/
        if (goflg){ /*freeze if stopped*/
          if (++nowtime==next[i].abstime){
             if (next[i].pattern==0xff)i=nowtime=0;
             PORTA=next[i++].pattern;
           }
        }

        /*added keyscan portion—1/4 scan each time*/
6       P1=scanpat[j]; /*uses top 4 bits,bottom 4 for input */
        newsc=(newsc<<4) | (P1 & 0xf);
        if(++j==4){
        j=0;
7       if (newsc!=oldsc) newkeyflg=1;
        scan=oldsc=newsc;
        newsc=0;
        }
   }

8  void solenoidupdate(void){
        if (next[entry].pattern & 1)
           display(92," on");
        else display(92,"off");
        if (next[entry].pattern & 2)
           display(81," on");
        else display(81,"off");
   }

9  void row2update(void){
        numdsp(65,entry);
        numdsp(71,next[entry].abstime);
        solenoidupdate();
   }
```

[5]This interrupt procedure is like that of Chapter 9 with the addition of the key scanning. With a single interrupt scheduler, the more that needs to be done, the fuller the interrupt becomes. To keep a balance, the keys are scanned 4 at a time so it takes four interrupts to finish the 4 × 4 scan of the keypad. That could help to balance the loading on the interrupts as opposed to doing all 4 scans once every 40mSec.

[6]The scan routine issues the scan pattern (one line of the four low) and reads back in the low part of the port.

[7]The flag signals when a change of state comes to the keypad. It is the shared signal to the keyinterp function that it can proceed.

[8]The solenoidupdate function simply sends "on" or "off" messages to the display beneath the "solenoid 1" and "solenoid 2" labels.

[9]Row2update refreshes the entire bottom row of the display when the information changes.

Figure 10–2b Continued

```c
void msec(uint x){
   uchar j;
   while (x-> 0){
      for (j=0;j</> 5;j++);
   }
}
void initialize(void){
   msecs = 0;
   TH0 = ~(8333/256);/*8333 at 10 MHz = .01 Sec*/
   TL0 = -8333%256;
   TMOD = 0x01;  /*mode 1-16 BIT*/
   TF0 = 0;
   TR0 = 1;   /*start timer 0 */
   ET0 = 1;   /*enable interrupt 0 */
   EA = 1;    /*global interrupt enable*/
   msec(15); /*set up LCD modes*/
   lcdcmd = 0x30; msec(4);
   lcdcmd = 0x30; msec(1);
   lcdcmd = 0x30; msec(1);
   lcdcmd = 0x38; msec(1); /*1/16 duty?*/
   lcdcmd = 0x0e; msec(1); /* dsp on*/
   lcdcmd = 0x01; msec(2);
   lcdcmd = 0x06; msec(1);
}

void main(void){
   pcmd=0x80; /*initialize solenoid I/O*/
   i=nowtime=newsc=newkeyflg=newnumbflg=goflg=0;
   for(i=0;i ;i++){
      next[i].abstime=0; next[i].pattern=0;
   }
   next[0].pattern=0xff;
   initialize();
   display(0,"entry");
   display(8,"time");
   display(15,"solenoid 1");
   display(26,"solenoid 2");
   forever{
      while (newkeyflg==0); /*wait for new key entry*/
      if (goflg==0){
         newkeyflg=0; /*reset for next new key*/
         if (scan==0x0000);/*0->key released*/
         else if (scanx0200){ /*0-9 key*/
            if (newnumbflg==0) {
               next[entry].abstime=0;
               newnumbflg=1;
            }
            for(x=0;scan & 1==0;scan>=1)x++;
```

The lines marked **10** are `display(0,"entry");` and following, and **11** marks `if (scan==0x0000);`

[10]These lines send the top row display information.

[11]The different user inputs are described in the text. The keyscan information is 16 bits with an increasing order assumed here. If the keypad were cross-wired, a translation table might be needed.

Figure 10–2b   Continued

```
               next[entry].abstime=10*next[entry].abstime+x;
               numdsp(71,next[entry].abstime);
           }
           else if (scan==0x0400){ /*"a" key*/
               if (entry ) entry++;
               newnumbflg=0;
               row2update();
           }
           else if (scan==0x0800){ /*"b" key*/
               newnumbflg=0;
               next[entry].pattern&=0x3;
               solenoidupdate();
           }
           else if (scan==0x1000); /*"c" key*/
           else if (scan==0x2000); /*"d" key*/
           else if (scan==0x4000){ /*"e" key*/
               if (entry) entry—;
               newnumbflg=0;
               row2update();
           }
       }
```
```
       if (scan==0x8000) /*"f" key*/
           goflg=~goflg;
       }
     }
   }
```

---

[12]It is much better to keep the entry in binary form as done here. To save individual decimal digits is inefficient and would have to be converted eventually for use by the scheduler.

[13]The test for the "F" key is independent of the g o f l g. Otherwise it would be impossible to get into the k e y i n t e r p portion to recognize the key. It would be a set up only once, run forever system.

Figure 10–2b Continued

```
$REGISTERBANK(0)
SOLENOIDII:DO;
$INCLUDE (REG51.DCL)
DECLARE AS LITERALLY 'LITERALLY',FOREVER AS 'WHILE 1',
    EXT AS 'EXTERNAL',EOM AS '0FFH';
DECLARE LCDDAT BYTE AT(8000H) AUX;
DECLARE LCDCMD BYTE AT(0C00RH) AUX;
DECLARE (NEWKEYFLG,NEWNUMBFLG,GOFLG) BIT;
DECLARE PCMD BYTE AT(2003H) AUX;
DECLARE PORTA BYTE AT(2000H) AUX;
DECLARE NEXT (20) STRUCTURE (ABSTIME WORD,PATTERN BYTE);
DECLARE (I,J,ENTRY,X) BYTE;
DECLARE (NOWTIME,SCAN,OLDSC,NEWSC) WORD;
```

[1]See most of the comments under the C version, since there are only small differences.

Figure 10–2c Solenoid Sequencer II—PL/M

```
DISPLAY:PROCEDURE (STARTLOC,SPTR);
DECLARE STARTLOC BYTE,SPTR WORD;
2  DECLARE (S BASED SPTR) BYTE CONSTANT;
   LCDCMD = 80H OR STARTLOC; CALL TIME(1);
   DO WHILE (S<>EOM);
      LCDDAT = S; CALL TIME(2);
      SPTR=SPTR+1;
   END;
END DISPLAY;

NUMDSP:PROCEDURE(NUMBER,DECIMALPT);
   /*THIS CONVERTS UP TO 16-BIT UNSIGNED BINARY
      VALUE TO STRING OF DISPLAYABLE ASCII DECIMAL
      DIGITS SENT TO THE LCD DISPLAY.  IT REPLACES
      LEADING ZEROS WITH BLANK SPACE. */
   DECLARE NUMBER WORD,DECIMALPT BIT;
   DECLARE M WORD, LDGZERO BIT;
   M = 10000;
   LDGZERO = 0;
   DO WHILE M>0;
      IF ((NUMBER/M)>0) OR (M = 1)
         OR ((M = 100) AND (DECIMALPT))
         THEN LDGZERO = 1;
      IF LDGZERO THEN LCDDAT=NUMBER/M+'0';
      ELSE LCDDAT = ' ';
      CALL TIME(1);
      NUMBER = NUMBER MOD M;
      IF (DECIMALPT) AND (M=100) THEN DO;
         LCDDAT = '.';
         CALL TIME(1);
      END;
      M = M/10;
   END;
END NUMDSP;

CYCLETIMER:PROCEDURE INTERRUPT 1 USING 1;
    /* REALTIME INTERRUPT ROUTINE EVERY 10 mSec*/
   DECLARE SCANPAT(*) BYTE CONSTANT (10H,20H,40H,80H);
   TH0 =HIGH(-8333);/*10mSec(10MHz OSC)*/
   TL0 =LOW(-8333);
   ET0 = 1;
    /*SOLENOID DRIVE PORTION*/
   IF GOFLG=1 THEN DO; /*FREEZE IF STOPPED*/
      NOWTIME=NOWTIME+1;
```

[2]It is necessary to set the display procedure to work out of constant space since the character string is put into constant space in the procedure call. It could be much more complicated if messages were sent from both CONSTANT, MAIN, and AUXILIARY space.

Figure 10–2c Continued

```
      IF NOWTIME=NEXT(I).ABSTIME THEN DO;
        IF (NEXT(I).PATTERN=0FFH) THEN DO;
          I,NOWTIME=0;
        END;
        PORTA=NEXT(I).PATTERN;
        I=I+1;
      END;
    END;
    /*ADDED KEYSCAN PORTION—1/4 SCAN EACH TIME*/
    P1=SCANPAT(I);
    NEWSC=ROL(NEWSC,4) OR (P1 AND 0FH);
    J=J+1;
    IF J=4 THEN DO;
      J=0;
      IF NEWSC<>OLDSC THEN NEWKEYFLG=1;
      SCAN,OLDSC=NEWSC;
      NEWSC=0;
    END;
END CYCLETIMER;


SOLENOIDUPDATE:PROCEDURE;
  IF (NEXT(ENTRY).PATTERN AND 01H)>0
    THEN CALL DISPLAY(92,.(' ON',EOM));
  ELSE CALL DISPLAY(92,.('OFF',EOM));
  IF (NEXT(ENTRY).PATTERN AND 02H)>0
    THEN CALL DISPLAY(81,.(' ON',EOM));
  ELSE CALL DISPLAY(81,.('OFF',EOM));
END SOLENOIDUPDATE;


ROW2UPDATE:PROCEDURE;
  CALL NUMBDSP(65,ENTRY);
  CALL NUMBDSP(71,NEXT(ENTRY).ABSTIME);
  CALL SOLENOIDUPDATE;
END ROW2UPDATE;


INITIALIZE: PROCEDURE;
  MSECS = 0;
  CALL TIME(150); /*SET OP LCD MODES*/
  LCDCMD = 30H; CALL TIME(41);
  LCDCMD = 30H; CALL TIME(2);
  LCDCMD = 30H; CALL TIME(2);
  LCDCMD = 38H; CALL TIME(2); /*1/16*/
  LCDCMD = 0EH; CALL TIME(2); /* DSP ON*/
  LCDCMD = 01H; CALL TIME(18);
  LCDCMD = 06H; CALL TIME(2);
```

Figure 10–2c   Continued

```
THO =HIGH(-8333); /*8333 AT 10 MHz = .01 Sec*/
TLO =LOW(-8333);
TMOD = 01H;   /*MODE 1-16 bit*/
TF0 = 0;
TR0 = 1;   /*START TIMER 0 */
ET0 = 1;   /*ENABLE INTERRUPT 0 */
EA = 1;    /*GLOBAL INTERRUPT ENABLE*/
END INITIALIZE;
I,NOWTIME,NEWSC=0; NEWKEYFLG,NEWNUMBFLG,GOFLG=0;
DO I=0 TO 20;
   NEXT(I).ABSTIME=0; NEXT(I).PATTERN=0;
END;
NEXT(0).PATTERN=0FFH;
CALL INITIALIZE;
CALL DISPLAY(0,.('ENTRY',EOM));
CALL DISPLAY(8,.('TIME',EOM));
CALL DISPLAY(15,.('SOLENOID 1',EOM));
CALL DISPLAY(26,.('SOLENOID 2',EOM));
DO FOREVER;
   DO WHILE NEWKEYFLG=0; /*WAIT FOR NEW KEY ENTRY*/
   END;
   NEWKEYFLG=0; /*RESET FOR NEXT NEW KEY*/
   IF GOFLG=0 THEN DO; /*ONLY WHEN STOPPED*/
      IF SCAN=0000000000000000B THEN; /*>-KEY RELEASED*/
      ELSE IF SCAN<10000000000B THEN DO; /*0-9 KEY*/
         IF NEWNUMBFLG=0 THEN DO;
            NEXT(ENTRY).ABSTIME=0;
            NEWNUMBFLG=1;
         END;
         X=0;
         DO WHILE (SCAN AND 1)=0;
            SCAN=SHR(SCAN,1);
            X=X+1;
         END;
         NEXT(ENTSRY).ABSTIME=10*NEXT(ENTRY).ABSTIME+X;
         CALL NUMBDSP(71,NEXT(ENTRY).ABSTIME);
      END;
      ELSE IF SCAN=0000010000000000B THEN DO; /*"A" KEY*/
         IF ENTRY<18 THEN ENTRY=ENTRY+1;
         NEWNUMBFLG=0;
         CALL ROW2UPDATE;
      END;
      ELSE IF SCAN=0000100000000000B THEN DO; /*"B" KEY*/
         NEWNUMBFLG=0;
         NEXT(ENTRY).PATTERN=(NEXT(ENTRY).PATTERN+1)AND 03H;
```

The 3 appears as a numbered callout beside `CALL DISPLAY(0,.('ENTRY',EOM));`

[3]Notice how the constant string is specified. In PL/M you must either add the e o m (end of message) code or define the first character as the length of the string if you are sending a string of varying length.

Figure 10–2c   Continued

```
      CALL SOLENOIDUPDATE;
      END;
      ELSE IF SCAN=0001000000000000B THEN; /*"C" KEY*/
      ELSE IF SCAN=0010000000000000B THEN; /*"D" KEY*/
      ELSE IF SCAN=0100000000000000B THEN DO; /*"E" KEY*/
        IF ENTRY >0 THEN ENTRY=ENTRY-1;
        NEWNUMBFLG=0;
        CALL ROW2UPDATE;
      END;
    END;
    IF SCAN=1000000000000000B THEN DO; /*"F" KEY*/
      GOFLG=NOT(GOFLG);
    END;
  END; /*END FOREVER*/
END SOLENOIDII;
```

Figure 10–2c Continued

The k e y s c a n task handles the reading of a user keypad. The keypad is a
4×4 matrix like the example from Chapter 3. The k e y s c a n task takes care of
the scanning at a 40mSec cycle time to eliminate bounce. When it is recognized
that a button has changed state, the value is forwarded to the k e y i n t e r p task.

The k e y i n t e r p task updates the solenoid sequence list. The list is shared
with the s o l e n o i d task, which operates much like that of Chapter 9. The information is also sent to the d i s p l a y task so the user can see the result of pushing
a button. Without a RTOS, "sending" is simply passing a parameter to the receiving routine (task). To simplify the example for the book, only the bare minimum
of user inputs and displays is used—user-friendly features should really be added
for a real system. For example, the solenoid sequence is not reorganized if times
are put in out of order. A 3-key sequence is demanded even if the first digits are
zero, and there is no backing up from a wrong entry. The task deciphers the k e y -
s c a n input to assemble a decimal number or to use 3 of the 6 alpha keys to carry
out predefined functions. The absolute time in decimal is entered numerically.
The "A" key means go on to the next time in the sequence. The "B" key increments the two states of the two solenoids (4 possibilities) with "on" and "off"
showing under the respective labels on the display. The "F" key toggles the
start/stop of the actual solenoid sequencing. Only when stopped is the user allowed to change the sequence table. Presumably custom key labels would be put
on the alpha keys of the keypad. This could be refined and made more friendly,
but the goal here is to illustrate the multitasking.

The d i s p l a y task assumes the LCD display module is initialized elsewhere or else does it when the task starts up. To avoid confilct with the
n u m d i s p l a y task, some resource management is used where the multitasking
system could allow both tasks to access the LCD at the same time.

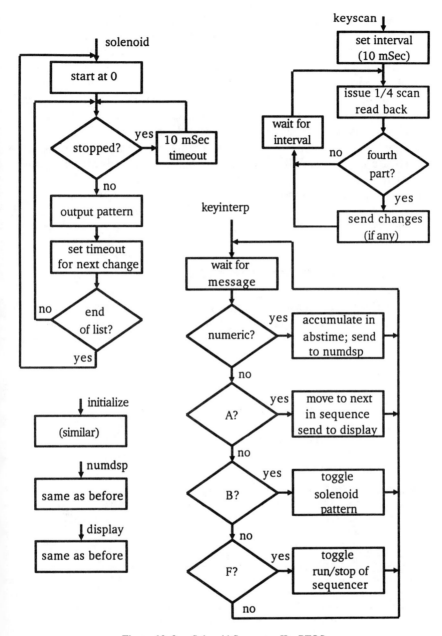

Figure 10–3a Solenoid Sequencer II—RTOS

The numdisplay task illustrates converting decimal to ASCII. The distinction between a routine and a task is a bit fuzzy until you move to an operating system—then there are distinct differences, and tasks can include several routines.

The hardware for this example, shown in Fig. 10–1a, is very standard. All the alphanumeric LCD modules seem to be based on the same controller chip, which makes the setup sequence identical. The sequence used in the examples is better understood from the data sheets; here it is just a "given." The modules have a 450nSec access time, which makes interface directly to a 12MHz 8051 somewhat marginal, but it seems that hundreds of projects have gotten by because it is a worst-case specification. If you don't want to drop back to 10MHz but want to stay strictly within specifications, it is possible to I/O drive the display by generating a write pulse through toggling a port bit in software. Well-written software will isolate the interface in a driver routine so, if the hardware changes, there is only one place needing the software change.

SOLENOID SEQUENCER II WITH RTOS

This is the first true *multi*tasking example in the book. This example is quite similar to the non-RTOS versions in partitioning. The task block diagram of Fig. 10–1b is an excellent way to visualize the interactions. The task flow charts are shown in Figure 10–3a while the acutal code is in Figures 10–3b through 10–3g. To show the variety of options within RTOS, the communication is done by a mix

---

```
1   #include <dcx2.h>
    #define pcmd XBYTE[0x2003]
    #define PORTA XBYTE[0x2000]
    struct {uint abstime;uchar pattern;} next [20] extern;
    bit goflg extern;

2   void solenoid(void) {
       pcmd=0x80; /*PORTA output*/
       forever{
          for (i=0;next[i].pattern!=0xff;i++){
```

[1]The example could have been done in separate files if desired. It is collected into three files to save space, but the approach I normally use is to put each task's object file into a library and then link all the modules together. The ITD and the task at a different priority are in different files because they are respectively in assembly language and in a different register bank which requires a different set of DCX interface routines. This is *multitasking* because the individual functions are all endless loops with system calls (the rqwait and rqsendmessage).

[2]Unlike the non-RTOS solutions, the key scan is a separate task from the solenoid updating. The operating system manages time events for separate tasks so they don't all have to fall in the same interrupt. This is a separate file because the header file is different for the different register banks and DCX relies on register banks for context switching. In the ITD you will see that the solenoid task is set at a different priority so it must be in a different bank.

Figure 10–3b   Solenoid Sequencer II—DCX and DCE

```
       if (goflg) { /*freeze count if stopped*/
          PORTA=next[i].pattern;
          rqwait(next[i+1].abstime-next[i].abstime,timeout);
       }
       else {
          rqwait(10,timeout); i--;
       }
    }
  }

*************************************************************

  #include <dcx1.h>
  #define lcddat XBYTE[0xc000]
  #define lcdcmd XBYTE[0x8000]

  void keyscan (void) {
     uchar code scanpat[]={0x10,0x20,0x40,0x80};
3    struct keyform {struct msghdr header;uint pat;}*keymsg;
     static uint oldsc; uint newsc; uchar j;
4    rqsetinterval(10); /*1/4 scan every 10 mSec*/
     forever{
        for(j=0,newsc=0;j<4;j++){
           P1=scanpat[j];
           newsc=(newsc<<4) | (P1 & 0xf);
           rqwait(0xff,interval);
        }
5       if (newsc!=oldsc){
6          keymsg=rqallocate(); /*report to keyinterp task*/
7          if (keymsg != 0){ /*if buffer free*/
8             keymsg->header.len=9;
              keymsg->header.nod=0;
              keymsg->header.flg=0;
              keymsg->header.cmd=0;
              keymsg->header.cmd=keycode;
              keymsg->header.sdt=
9                     rqtaskid<4 | taskof(0x86);
              keymsg->pat=newsc;
```

[3]The messages in this example are sent in on-chip RAM which allows eventual sending of messages to other nodes (other boards). The header, not described, inlcudes the address and length information. It is based so it can be located wherever the operating system assigns it space. The "real" information in the message is only the 2 bytes that are the keyscan result.

[4]The interval is set at 10 to give a full scan every 40mSec. That ought to eliminate bounce problems while still giving fast enough response for the user.

[5]Messages are sent only if the keyscan information has changed. Otherwise a new message every 40mSec could exhaust the supply of message buffers quite quickly if the receiving task didn't free them up.

[6]The allocate obtains an "envelope" buffer to hold the message.

[7]If no envelope is available, no message can be sent.

[8]These steps fill the message header and the message data itself. The 20-byte limit allows only 13 bytes of actual data in each message.

[9]It is necessary to "look up" the current number of the desired task (called task id) by using its unchanging number set in the ITD (called its function id). The t a s k o f function, hidden in the header, handles this.

Figure 10–3b Continued

```
10                    rqsendmessage(keymsg);
                      }
                 oldsc=newsc;
              }
           }
        }

   void display (void){
      struct {struct msghdr header;
                    uchar loc;uchar ch[12];}*dmsg;
      uchar i;
      forever{
11       dmsg=rqwaitm(0xff,message);
         lcdcmd = 0x80 | dmsg->loc; /*start location*/
         rqwait(1,timeout);
         for (i=0;dmsg->ch[i]!=0;i++){
            lcddat = dmsg->ch[i];
            rqwait(1,timeout);
         }
         dmsg->hdr.flg = 0x80; /*reply msg*/
12       rqsendmessage(dmsg);
      }
   }


   void numdisplay(void){
      struct {struct msghdr header;
                    uchar loc;uint dat;}*nmsg;
      uint number;
      forever{
11       nmsg=rqwaitm(0xff,message);
         lcdcmd = 0x80 | nmsg->loc;
         rqwait(1,timeout);
         number=nmsg->dat;
         uint m = 10000;
         bit ldgzero = 0;           do{
            if ((((number/m)>0) || (m == 1))
               || ((m == 100) && (dpt == 1)))
               ldgzero = 1;
            if (ldgzero == 1)
```

[10]Here the message is sent. The destination task (and node) is specified in the header portion of the message. It *is* also possible to send messages located in off-chip RAM, but then there is no allocation of space—each message occupies a fixed space and is passed only by exchanging the pointer with another task.

[11]With no task signaling mechanism, DCX and DCE here rely on a reply message as the indication that the display write task is done. This is also one way to avoid two tasks addressing the LCD at the same time (which would mess up the cursor locations).

[12]Converting the message to a reply by setting the msb of flag high will allow the task which sent the message to know it is done. If the message might have come from another board, it would be better to "OR" the flag with 80 to preserve any nonzero bits of the flag.

Figure 10–3b   Continued

```
        lcddat = number/m+'0';
     else lcddat = ' ';
     rqwait(1,timeout);
     number %= m;
     if (m == 100) {
        lcddat = '.';
        rqwait(1,timeout);
     }
   }
     while((m /= 10)>0);
     nmsg->flg=0x80; /*send reply*/
12   rqsendmessage(nmsg);
   }
 }

/*following 4 functions are not tasks-go with keyinterp*/

13  void senddisplay (uchar loc;uchar *msg;uchar datatype){
    struct {struct msghdr header;
                  uchar loc;uchar ch[12];}*sdmsg;
    msg=rqallocate();
    if (msg>0){ /*buffer available*/
      sdmsg->hdr.len=20;
      sdmsg->hdr.nod=0;
      sdmsg->hdr.flg=0;
      sdmsg->hdr.cmd=0;
      sdmsg->hdr.sdt=rqtaskid<<4 | taskof(0x82);
      sdmsg->loc=loc;
      for (i=0;*msg!=0;i++) sdmsg->ch[i]=*msg++;
      rqsendmessage(sdmsg); /*send to display task*/
      rqwait(0xff,message);
    }
  }

void sendnumb (uchar loc;uint number){
   struct {struct msghdr header;
                 uchar loc;uint number;}*snmsg;
   snmsg=rqallocate();
   if (snmsg>0){ /*buffer available*/
     snmsg->hdr.len=10;
     snmsg->hdr.nod=0;
     snmsg->hdr.flg=0;
     snmsg->hdr.sdt=rqtaskid<<4 | taskof(0x84);
     snmsg->hdr.cmd=0;
     snmsg->loc=loc;
     snmsg->number=number;
     rqsendmessage(snmsg); /*send to numdsp task*/
     rqwait(0xff,message);
   }
 }
```

[13]These were collected into functions (not tasks) to simplify what is a repeated operation.

Figure 10–3b Continued

```
14 void solenoidupdate(void){
     if (next[entry].pattern & 1)
        senddisplay(92," on");
     else senddisplay(92,"off");
     if (next[entry].pattern & 2)
        senddisplay(81," on");
     else senddisplay(81,"off");
   }

15 void row2update(void){
     sendnumb(65,entry);
     sendnumb(71,next[entry].abstime);
     solenoidupdate();
   }

   void keyinterp (void){
     struct msgform {struct msghdr header;uint scan;}*msg;
     static uchar entry;
     uchar x;
     uint scan;
     forever{
16     msg=rqwaitm(0xff,message);
       scan=msg->scan; /*new key condition to process*/
17     rqdeallocate(msg);
       if (goflg==0){    /*update only if stopped*/
         if (scan==0x0000);/*0->key released*/
18       else if (scan<0x0200){ /*0-9 key*/
           if (newnumbflg==0) {
             next[entry].abstime=0;
             newnumbflg=1;
           }
           for(x=0;scan & 1==0;scan>>=1)x++;
           next[entry].abstime=10*next[entry].abstime+x;
           sendnumb(71,next[entry].abstime);
         }
         else if (scan==0x0400){ /*"a" key*/
           if (entry<18) entry++;
           newnumbflg=0;
           row2update();
```

[14]This function sends the "on" and "off" to the bottom row under "solenoid 1" and "solenoid 2."
[15]This function fills the entire bottom display row with the most recent information.
[16]This wait is without any timeout (0xff). The pointer to the message is returned from the system call. The data in the message must be used or transferred before the buffer is released and the sending task must not do anything to the buffer once it is sent.
[17]Deallocating is necessary if the system is to not crash due to lack of buffers. The buffers are all on-chip and there are probably no more than 5 or 6 of them (20 bytes long by default).
[18]The key codes and the interpretation are described in the first non-RTOS version and are not changed.

Figure 10–3b   Continued

```
         }
         else if (scan==0x0800){ /*"b" key*/
            newnumbflg=0;
            next[entry].pattern&=0x3;
            solenoidupdate();
         }
         else if (scan==0x1000); /*"c" key*/
         else if (scan==0x2000); /*"d" key*/
         else if (scan==0x4000){ /*"e" key*/
            if (entry) entry--;
            newnumbflg=0;
            row2update();
         }
      }
      if (scan==0x8000){ /*"f" key*/
         goflg=~goflg;
      }
   }
}


19 void initialize(void){
      nowtime=newsc=goflg=0;
      rqwait(15,timeout); /*set up LCD modes*/
20    lcdcmd = 0x30; rqwait(4,timeout);
      lcdcmd = 0x30; rqwait(1,timeout);
      lcdcmd = 0x30; rqwait(1,timeout);
      lcdcmd = 0x38; rqwait(1,timeout); /*1/16 duty?*/
      lcdcmd = 0x0e; rqwait(1,timeout); /* dsp on*/
      lcdcmd = 0x01; rqwait(2,timeout);
      lcdcmd = 0x06; rqwait(1,timeout);
      for(i=0;i<20;i++){
         next[i].abstime=next[i].pattern=0;
      }
      next[0].pattern=0xff;
21    senddisplay(0,"entry");
      senddisplay(8,"time");
      senddisplay(15,"solenoid 1");
      senddisplay(26,"solenoid 2");
```

[19]The initialize task avoids any premature use of the LCD by being the only running task at start-up. When it is finished it creates (puts on record with the operating system) the other tasks.

[20]The initialization of the LCD must happen first. The details are the same as the previous (non-RTOS) versions except the system timer is used with the rqwait to get the delays between writes to the LCD.

[21]These lines set up the first line display and are sent out only once.

Figure 10–3b Continued

```
22    rqcreatetask(0x800e); /*solenoid*/
      rqcreatetask(0x801d); /*numdisplay*/
      rqcreatetask(0x802c); /*keyscan*/
      rqcreatetask(0x803b); /*keyinterp*/
23    rqdeletetask(rqtaskid); /*deletes itself*/
      }

      ****************************************************

      $INCLUDE(RMX51A.MAC)
      EXTRN CODE (solenoid,display,initialize)
      EXTRN CODE (numdisplay,keyscan,keyinterp)
      CSEG AT 0FFF0H
      ITD1:
      %ITD(ENTRY1,10,81H,101B,2,0,ITD2)
      ENTRY1:LJMP initialize
      CSEG AT 8000H
      ITD2:
24    %ITD(ENTRY1,10,82H,101B,2,0,ITD2+1)
      ENTRY2:LJMP display
      ITD3:
25    %ITD(ENTRY1,10,83H,110B,3,0,ITD4)
      ENTRY3:LJMP solenoid
      ITD4:
      %ITD(ENTRY1,10,84H,101B,2,0,ITD5)
      ENTRY4:LJMP numdisplay
      ITD5:
      %ITD(ENTRY1,10,85H,101B,2,0,ITD6)
      ENTRY5:LJMP keyscan
      ITD6:
      %ITD(ENTRY1,10,86H,101B,2,0,ITD6+1)
      ENTRY6:LJMP keyinterp
  END
```

---

[22]These tasks could have been created at power-up if all the ITDs had been chained together, but it was desired to keep the other LCD access from beginning before the top line was written. Only the initialize and display tasks start up at the beginning because the ITD chain stops after the second task (ITD2+1).

[23]Since it is no longer needed, the initialize task deletes itself.

[24]Notice that the start-up chaining stops with the second task.

[25]This task is a different priority so it must use a different register bank (110=rb2 rather than 101=rb2).

Figure 10–3b　Continued

```
#include  <rtxtiny.h>
#define lcddat XBYTE[0xc000]
#define lcdcmd XBYTE[0x8000]
#define pcmd XBYTE[0x2003]
#define PORTA XBYTE[0x2000]
struct {uint abstime;uchar pattern;} next [20] extern;
uint numb;
uchar i,status,scan;
bit goflg,lcdbusy;

void interval(void) _task_ 3{
   for(;;);
1       os_wait(K_TMO,100);
2       os_send_signal(2);
   }
}


void solenoid(void) _task_ 3 {
   for(;;){
      for (i=0;next[i].pattern!=0xff;i++){
         if (goflg){ /*freeze count if stopped*/
            PORTA=next[i].pattern;
3            os_wait(K_TMO,next[i+1].abstime-next[i].abstime);
         }
         else {
4            os_wait(K_TMO,10); i-;
         }
      }
   }
}


void keyscan (void)_task_ 4 {
   uchar code scanpat[]={0x10,0x20,0x40,0x80};
   static uint oldsc; uint newsc; uchar j;
   forever{
      for(j=0,newsc=0;j<4;j++){
      P1=scanpat[j];
      newsc=(newsc<<4) | (P1 & 0xf);
      os_wait(K_TMO,10);
   }
```

[1]The interval must be created by a separate task in RTXtiny. Here it makes a delay of 100 ticks (presumably 1mSec intervals).

[2]Here the interval event is sent to the t i m e t a s k (#2). Note that interval doesn't relinquish the processor until it loops back to o s_w a i t.

[3]The solenoid interval runs as a timeout for the time until the next state change.

[4]The task must go into a wait to avoid hogging the processor. Otherwise no other tasks could run and the g o f l g value could never be changed. This is called "deadlock."

Figure 10–3c   Solenoid Sequencer II—RTXtiny

```
         if (newsc!=oldsc){
           /*report to keyinterp task*/
           scan=newsc; /*put in public variable*/
5          os_send_signal(5); /*to keyinterp*/
         }
         oldsc=newsc;
       }
     }
   }

   void numdisplay(void) _task_ 2 {
     uint number;
     forever{
       while (os_wait(K_SIG,0xff)==K_TMO);
       while (lcdbusy) os_wait(K_TMO,10);
       lcdbusy=1;
       lcdcmd = 0x80 | loc; /*start location*/
       number=value;
       uint m = 10000;
       bit ldgzero = 0;
       do{
          if ((((number/m)>0) || (m == 1))
             || ((m == 100) && (dpt == 1)))
             ldgzero = 1;
          if (ldgzero == 1)
             lcddat = number/m+'0';
          else lcddat = ' ';
          os_wait(K_TMO,1);
          number %= m;
          if (m == 100) {
             lcddat = '.';
             os_wait(K_TMO,1);
          }
       }
       while((m /= 10)>0);
       lcdbusy=0;
     }
   }

   void display (void) _task_ 1 {
     forever{
6      while (os_wait(K_SIG,0xff)==K_TMO);
```

[5]The messages are passed by setting a shared variable and sending a signal to awaken the receiving task.

[6]The task communication is by the signal. It is not clear how to request a non-timeout wait for a signal—hopefully 0xff or 0 will give an endless wait for a signal. If you cannot wait indefinitely, here the wait is for as long as possible (0xff) in hopes that the signal will have arrived. Some caution is necessary since only one signal bit exists for the task.

Figure 10–3c Continued

```
7        while (lcdbusy) os_wait(K_TMO,10);
         lcdbusy=1;
         lcdcmd = 0x80 | loc; /*start location*/
         os_wait(K_TMO,1);
8        while(*dsp){
             lcddat = *dsp++;
             os_wait(K_TMO,2);
         }
9        lcdbusy=0;
     }
 }
     /*following 2 functions are not tasks-go with keyinterp*/

10   void solenoidupdate(void){
         do os_wait(0) while (lcdbusy);
         loc=81;
         if (next[entry].pattern & 1) dsp=&(" on");
         else dsp=&("off");
         os_send_signal(1);
         do os_wait(0) while (lcdbusy);
         loc=92;
         if (next[entry].pattern & 2) dsp=&(" on");
         else dsp=&("off");
         os_send_signal(1); os_wait(0);
     }

 void row2update(void){
     do os_wait(0) while (lcdbusy);
     loc=65; numb=entry;
     os_send_signal(2);
     do os_wait(0) while (lcdbusy);
     loc=71; numb=next[entry].abstime;
```

[7]The resource management is handled by this shared flag. Once a task starts using the LCD, the other tasks should not write because the cursor is set to the wrong place. Setting lcdbusy claims the LCD for this task. Without preemption, the system must do a timeout to allow the display task to start and then wait until the lcdbusy flag is cleared. If it did a while loop without the timeout, it would fail to relinquish control to the task which could clear the flag. If it went on without waiting for the display task to finish, it would overwrite the display pointer and mess up the message.

[8]With no message capability (unlike the larger RTX), messages are handled by a signal and a shared pointer (here dsp). Since dsp is purposely a generic pointer to void, all compiler complaints about which type of memory and the way it is interpreted should be silenced. Franklin supports 3-byte pointers which have a third byte identifying which memory space is being pointed to. In this case it will be an array (a string) in code space.

[9]The task must release the resource when done or nothing else can use it.

[10]These two functions are not tasks, but they group the display write functions. Solenoidupdate sends "on" and "off" under the "solenoid 1" and "solenoid 2" headings on the 2-line display. A note of caution: I believe the sending of a signal does not stop the task. lcdbusy flag is used to keep the task from sending a second message before the first is processed. Without preemption, the sending task must let the other tasks go ahead with an os_wait for time 0 or 1 (some systems call this a rescheduling request).

Figure 10–3c Continued

```
        os_send_signal(2); os_wait(0);
        solenoidupdate();
    }
11  void keyinterp(void) _task_ 5 {
        static uchar entry;
        bit newnumbflg=0;
        uchar x;
        forever{
            os_wait(K_SIG,0xff); /*signal from keyscan*/
            if (goflg==0){      /*update only if stopped*/
                if (scan==0x0000);/*0->key released*/
                else if (scan<x0200){ /*0-9 key*/
                    if (newnumbflg==0) {
                        next[entry].abstime=0;
                        newnumbflg=1;
                    }
                    for(x=0;scan & 1==0;scan>>=1)x++;
                    next[entry].abstime=10*next[entry].abstime+x;
                    loc=71;numb=next[entry].abstime; os_send_signal(2);
                }
                else if (scan==0x0400){ /*"a" key*/
                    if (entry<18) entry++;
                    newnumbflg=0;
                    row2update();
                }
                else if (scan==0x0800){ /*"b" key*/
                    newnumbflg=0;
                    next[entry].pattern&=0x3;
                    solenoidupdate();
                }
                else if (scan==0x1000); /*"c" key*/
                else if (scan==0x2000); /*"d" key*/
                else if (scan==0x4000){ /*"e" key*/
                    if (entry) entry--;
                    newnumbflg=0;
                    row2update();
                }
            }
            if (scan==0x8000){ /*"f" key*/
                goflg=~goflg;
            }
        }
    }

void initialize(void) _task_ 0{
    pcmd=0x80; /*initialize solenoid I/O*/
    nowtime=newsc=goflg=0;
```

[11]Other than waiting for a signal rather than a message, this task is the same as the previous (DCX) example.

Figure 10–3c Continued

```
12    os_wait(K_TMO,15); /*set up LCD modes*/
      lcdcmd = 0x30; os_wait(K_TMO,4);
      lcdcmd = 0x30; os_wait(K_TMO,2);
      lcdcmd = 0x30; os_wait(K_TMO,2);
      lcdcmd = 0x38; os_wait(K_TMO,2); /*1/16 duty?*/
      lcdcmd = 0x0e; os_wait(K_TMO,1); /* dsp on*/
      lcdcmd = 0x01; os_wait(K_TMO,2);
      lcdcmd = 0x06; os_wait(K_TMO,1);
      for(i=0;i<20;i++){
         next[i].abstime=next[i].pattern=0;
      }
      lcdbusy=0;
      next[0].pattern=0xff;
      os_create(1); /*display*/
13    loc=0;dsp=&"entry";
      os_send_signal(1);
      do os_wait(0) while (lcdbusy);
      loc=8;dsp=&"time";
      os_send_signal(1);
      do os_wait(0) while (lcdbusy);
      loc=15;dsp=&"solenoid 1";
      os_send_signal(1);
      do os_wait(0) while (lcdbusy);
      loc=26;dsp=&"solenoid 2";
      os_send_signal(1); os_wait(0);
14    os_create(2); /*numdisplay*/
      os_create(3); /*solenoid*/
      os_create(4); /*keyscan*/
      os_create(5); /*keyinterp*/
15    os_delete(0); /*deletes itself*/
   }
```

---

[12]The initialization of the LCD must happen first. The details are the same as the previous versions except the system timer is used with the os_wait to get the delays between writes to the LCD. It is convenient to delay creation of the other tasks to finish the LCD initialization without conflict.

[13]The display messages are not sent, but rather a signal is sent indicating that the display data is ready.

[14]RTXtiny requires the sucessive tasks be created by the first task (0) or its followers.

[15]The initialization task is no longer needed so it can be deleted.

Figure 10–3c Continued

of signaling and messages where possible. In certain cases, the example employs protected resource management so the LCD is not addressed by two tasks at once. The examples follow much the same form as the previous ones with the replacement of parameter passing with messages and signaling. As the examples grow in complexity, you begin to see the advantage of multitasking as well as the reasons to make "generic" driver tasks that can be reused.

```
   #include <usxcfg.h>
   #include <usx.h>
   #define forever for(;;)
1  #define uchar unsigned char
   #define uint unsigned
   #define tenmsecs 1
   #define keymbx 1
   #define dspmbx 2
   #define nummbx 3
   #define lcddat XBYTE[0xc000]
   #define lcdcmd XBYTE[0x8000]
   #define pcmd XBYTE[0x2003]
   #define PORTA XBYTE[0x2000]
   bit goflg;
   struct {uint abstime;uchar pattern;} next [20];
   uchar entry;
   uint status,t;
   uchar slotini,slotdsp,slotnum,slotsol,slotks,slotki;
   uchar xdata free_memory[MEM_ALLOCATION];

   void solenoid(void){
      uchar code scanpat[]={0x10,0x20,0x40,0x80};
       /*solenoid drive portion*/
      uint nowtime;
      period(tenmsecs,1);
      forever{
2        wteset(tenmsecs,0);
         decevt(tenmsecs);
         if (goflg){ /*freeze if stopped*/
            if (++nowtime==next[i].abstime){
               if (next[i].pattern==0xff)i=nowtime=0;
               PORTA=next[i++].pattern;
               }
            }
            /*added keyscan portion-1/4 scan each time*/
3        P1=scanpat[j];
         newsc=(newsc<<4) | (P1 & 0xf);
         if(++j==4){
            j=0;
            scan=oldsc=newsc;
```

[1]Labels help make more sense out of system calls.

[2]This models the non-RTOS version with a regular 10mSec event and a counter. The approach could be rather to do a simple wait for the computed time until the next state change for the solenoid. Depending on the overhead for context switching, the latter might be more efficient for only the solenoids, but by combining the keyscan in the same task, the one event does double duty.

[3]The scan is done in four separate parts at a 10mSec rate to even up the loading on the individual events.

Figure 10–3d   Solenoid Sequencer II—USX

```
4       if (newsc!=oldsc) sndmsg(keymbx,&scan,5);
        newsc=0;
      }
    }
  }

  void display (void){
    struct {uchar loc;uchar ch[12];}*dmsg;
    uchar i;
5   lcdcmd = 0x30; for(t=0;t<400;t++);
    lcdcmd = 0x30; for(t=0;t<400;t++);
    lcdcmd = 0x30; for(t=0;t<400;t++);
    lcdcmd = 0x38; for(t=0;t<400;t++);
    lcdcmd = 0x0e; for(t=0;t<20;t++);
    lcdcmd = 0x01; for(t=0;t<20;t++);
    lcdcmd = 0x06; for(t=0;t<20;t++);
    forever{
6     dmsg=rcvmsg(dspmbx,0);
7     getres(lcd,0);
      lcdcmd = 0x80 | dmsg->loc; /*start location*/
      for(t=0;t<20;t++);
      for (i=0;dmsg->ch[i]!=0;i++){
        lcddat = dmsg->ch[i];
        for(t=0;t<20;t++);
      }
8     relres(lcd);
    }
  }

  void numdisplay(void){
    struct {uchar loc;uint dat;}*nmsg;
    uint number;
    forever{
      nmsg=rcvmsg(mummbx.0);
      getres(lcd,0);
      lcdcmd = 0x80 | nmsg->loc;
      for(t=0;t<20;t++);
      number=nmsg->dat;
      uint m = 10000;
      bit ldgzero = 0;
      do{
```

[4]The message holding the changed scan result is sent to the **keyinterp** task. It is given an arbitrary priority as a message.

[5]This is the setup of the LCD module. The details are derived from the data sheets and are of no paticular interest here.

[6]The task waits forever (0) for a message in its mailbox. The returned result is the pointer to the message—the message probably sits where the sending task put it and should not be overwritten by the sending task before the receiver is finished with it. The part ahead of the forever is the one-time initialization of the LCD.

[7]This resource manager claims the exclusive use of the LCD. There is nothing inherent to stop another task from use of it except good programming practice.

[8]This releases the LCD for some other task's use.

Figure 10–3d Continued

```
        if ((((number/m)>0) || (m == 1))
            || ((m == 100) && (dpt == 1)))
            ldgzero = 1;
        if (ldgzero == 1)
            lcddat = number/m+'0';
        else lcddat = ' ';
        for(t=0;t<20;t++);
        number %= m;
        if (m == 100) {
            lcddat = '.';
            for(t=0;t<20;t++);
        }
    }
    while((m /= 10)>0);
    relres(lcd);
  }
}

/*following 2 functions are not tasks—go with keyinterp*/

void solenoidupdate(void){
  if (next[entry].pattern & 1)
    sndmsg(dspmbx,&(92," on"),5);
  else sndmsg(dspmbx,&(92,"off"),5);
  if (next[entry].pattern & 2)
    sndmsg(dspmbx,&(81," on"),5);
  else sndmsg(dspmbx,&(81,"off"),5);
}

void row2update(void){
  struct {uchar loc,uint numb} envelope;
  envelope.loc=65;
  envelope.numb=entry;
  sndmsg(nummbx,&envelope,5);
  numdsp(71,next[entry].abstime);
  envelope.loc=71;
  envelope.numb=next[entry].abstime;
  sndmsg(nummbx,&envelope,5);
  solenoidupdate();
}

void keyinterp (void){
  uint *msg;
  uchar x;
  uint scan;
  bit newnumbflg;
  sndmsg(dspmbx,&(0,"entry"),5);
  sndmsg(dspmbx,&(8,"time"),5);
```

Figure 10–3d   Continued

```
     sndmsg(dspmbx,&(15,"solenoid 1"),5);
     sndmsg(dspmbx,&(26,"solenoid 2"),5);
     forever{
9        msg=rcvmsg(keymbx,0);
         scan=msg; /*new key condition to process*/
         if (goflg==0){    /*update only if stopped*/
           if (scan==0x0000);/*0->key released*/
10         else if (scan<0x0200){ /*0-9 key*/
                if (newnumbflg==0) {
                next[entry].abstime=0;
                newnumbflg=1;
                }
                for(x=0;scan & 1==0;scan>>=1)x++;
                next[entry].abstime=10*next[entry].abstime+x;
                sendnumb(71,next[entry].abstime);
           }
           else if (scan==0x0400){ /*"a" key*/
                if (entry<18) entry++;
                newnumbflg=0;
                row2update();
           }
           else if (scan==0x0800){ /*"b" key*/
                newnumbflg=0;
                next[entry].pattern&=0x3;
                solenoidupdate();
           }
           else if (scan==0x1000); /*"c" key*/
           else if (scan==0x2000); /*"d" key*/
           else if (scan==0x4000){ /*"e" key*/
                if (entry) entry-;
                newnumbflg=0;
                row2update();
           }
         }
11       if (scan==0x8000){ /*"f" key*/
           goflg=~goflg;
         }
       }
     }
   void main(void{
     MEM_DEF *memptr;
     usxini();
```

[9]The task waits forever (0) for a message from the keypad.

[10]The first ten bits represent the 0–9 keys. The numeric conversion is carried out by multiplying the previous entries by ten since each additional key indicates that the previous keys were one place over from the previous knowledge.

[11]The "F" key must be interpreted outside the g o f l g test because otherwise the flag would block ever changing the flag back.

Figure 10–3d   Continued

```
      usrclk_init();
      memptr=(MEM_DEF *)free_memory;
      memptr->mem_size=MEM_ALLOCATION-sizeof(struct mem_def);
      memptr->task_slt=SYSTEM;
      memptr->mem_nxt=NULL;
      relmem((usxmem_t *)(++memptr));
      pcmd=0x80; /*initialize solenoid I/O*/
      nowtime=newsc=goflg=0;
      for(i=0;i<20;i++){
         next[i].abstime=next[i].pattern=0;
      }
      next[0].pattern=0xff;
      slotdsp=runtsk(5,&display,20);
      slotnum=runtsk(5,&numdisplay,20);
      slotsol=runtsk(10,&solenoid,20);
      slotks=runtsk(5,&keyscan,20);
      slotki=runtsk(5,&keyinterp,20);
      begusx();
   }
```

Figure 10–3d   Continued

```
   #include <CMXFUNC.h>
   #define uchar unsigned char
   #define uint unsigned
   #define lcddat XBYTE[0xc000]
   #define lcdcmd XBYTE[0x8000]
   #define pcmd XBYTE[0x2003]
   #define PORTA XBYTE[0x2000]
   uchar slotini,slotdsp,slotnum,slotsol,slotks,slotki;
   uchar LCD;
   bit goflg;
1  struct {uint abstime;uchar pattern;} next [20];
   uchar entry; uint status,t;

2  void solenoid(void){ /*solenoid drive*/
      uchar i;
      forever{
         for (i=0;next[i]!= 0xff;i++){
            if (goflg && next[i]!=0xff){ /*freeze if stopped*/
3              cxtwatm(next[i+1].pattern-next[i].pattern);
               PORTA=next[i++].pattern;
            }
```

[1]This is a more flexible form of schedule—no predefined sequence as was done in Chapter 9.
[2]This task does the updating of the solenoid states.
[3]The wait is for a time—here the time until the next state change.

Figure 10–3e   Solenoid Sequencer II—CMX

```
4                cxtwatm(100); i--;
             }
          }
       }
    }

    void keyscan (void){
       uchar code scanpat[]={0x10,0x20,0x40,0x80};
       static uint oldsc; uint newsc; uchar j;
       forever{
          for(j=0,newsc=0;j<4;j++){
             P1=scanpat[j];
             newsc=(newsc<<4) | (P1 & 0xf);
             cxtwatm(1); /*10 mSec interval*/
          }
5         if (newsc!=oldsc) cxmssend(slotkey,&newsc);
          oldsc=newsc;
       }
    }

    void display (void){
       struct {uchar cursor;uchar ch[19];}msg;
       for(t=0;t<400;t++); /*set up LCD modes*/
6      lcdcmd = 0x30; for(t=0;t<400;t++);
       lcdcmd = 0x30; for(t=0;t<400;t++);
       lcdcmd = 0x30; for(t=0;t<400;t++);
       lcdcmd = 0x38; for(t=0;t<400;t++); /*1/16 duty?*/
       lcdcmd = 0x0e; for(t=0;t<20;t++); /* dsp on*/
       lcdcmd = 0x01; for(t=0;t<20;t++);
       lcdcmd = 0x06; for(t=0;t<20;t++);
       forever{
7         cxmswatm(&msg,0); /*wait for display msg*/
8         cxrsrsv(LCD);
          lcdcmd = 0x80 | msg.cursor; /*start location*/
```

[4]If the system is stopped, this task must still "get out of the way" so other tasks can run. Otherwise the goflg could never be changed by another task and the system would be "deadlocked."

[5]The sending of a message requires the destination task's slot and the address of the message information. The message information is actually transferred to a buffer at the receiving task so the sending task can reuse the information space right away.

[6]These lines initialize the LCD display module. They use a simple "burn time" for loop because the smallest system delay unit is one "tick" which is assumed to be 10mSec—much longer than is required for most display update commands.

[7]This wait for a message is forever. (0). The display task will not wake up until a message arrives. The message is actually filled in by the operating system from the task's message buffers so the definition at the top of this task sets actual space rather than a pointer to space.

[8]This call reserves the LCD resource so no other task (esp numdisplay) addresses the display once the cursor is set. There may be alternate ways to ensure this as discussed in note 11, but this has a certain foolproof sense to it. The only hazard is the possibility of deadlock—if the resource is unavailable and a high-priority task does not sleep for a while if the resource is unavailable, it could keep the other task from finishing and releasing the resource.

Figure 10–3e Continued

```
        for(t=0;t<20;t++);
        for (i=0;i<19 && msg.ch[i]){
            lcddat = msg.ch[i];
            for(t=0;t<20;t++);
        }
9       cxrsrel(LCD);
    }
}

void numdisplay(void){
    uint number;
    forever{
        cxmswatm(&number,0); /*wait for msg*/
        cxrsrsv(LCD);
        uint m = 10000;
        bit ldgzero = 0;
        do{
            if ((((number/m)>0) || (m == 1))
                || ((m == 100) && (dpt == 1)))
                ldgzero = 1;
            if (ldgzero == 1)
                lcddat = number/m+'0';
            else lcddat = ' ';
            for(t=0;t<400;t++);
            number %= m;
            if (m == 100) {
                lcddat = '.';
                for(t=0;t<20;t++);
            }
        }
        while((m /= 10)>0);
        cxrsrel(LCD);
    }
}

/*following 2 functions are not tasks—go with keyinterp*/

void solenoidupdate(void){
    if (next[entry].pattern & 1)
10      cxmssend(slotdsp,&(92," on"));
    else cxmssend(slotdsp,&(92,"off"));
    if (next[entry].pattern & 2)
        cxmssend(slotdsp,&(81," on"));
    else cxmssend(slotdsp,&(81,"off"),5);
}

void row2update(void){
```

[9] Be sure to *release* the resource when finished to avoid deadlock within your own task.

[10] The sending of a message requires the task to which it is sent and the pointer to the data to send. The information is transferred to a dedicated buffer at the receiving task. Here the information is the starting cursor address and the display characters.

Figure 10–3e   Continued

```
    struct {uchar loc,uint numb} msg;
    msg.loc=65; msg.numb=entry;
    cxmssend(slotnum,&msg);
    numdsp(71,next[entry].abstime);
    msg.loc=71; msg.numb=next[entry].abstime;
    cxmssend(slotnum,&msg);
    solenoidupdate();
  }

void keyinterp (void){
    uchar x;
    uint scan;
    bit newnumbflg;
11  cxmssend(slotdsp,&(0,"entry"));
    cxmssend(slotdsp,&(8,"time"));
    cxmssend(slotdsp,&(15,"solenoid 1"));
    cxmssend(slotdsp,&(26,"solenoid 2"));
    forever{
12      cxmswatm(&scan,0); /*wait forever*/
          /*new key condition to process*/
        if (goflg==0){    /*update only if stopped*/
        if (scan==0x0000);/*0->key released*/
        else if (scan<0x0200){ /*0-9 key*/
            if (newnumbflg==0) {
              next[entry].abstime=0;
              newnumbflg=1;
            }
            for(x=0;scan & 1==0;scan>>=1)x++;
            next[entry].abstime=10*next[entry].abstime+x;
            sendnumb(71,next[entry].abstime);
        }
        else if (scan==0x0400){ /*"a" key*/
            if (entry<18) entry++;
            newnumbflg=0;
            row2update();
        }
        else if (scan==0x0800){ /*"b" key*/
            newnumbflg=0;
```

[11]Unlike some earlier systems, messages queue up for the task so it is not strictly necessary to let the display task begin running before going on to send the next message. There are also message sending calls that transfer to the receiving task and require a reply signal before the sending task proceeds. That could allow the task that sent the message to sleep until the display activity is done. As long as the display task is higher priority than the tasks sending the messages, it should be possible to keep other tasks from sending additional messages except for the fact that the display task waits between characters, which would release the processor to some other task such as numdisplay, which would also address display. Needless to say, a thorough study of the instruction manual is recommended!

[12]The wait for a message is forever (0) because the task has nothing else to do until a new key arrives. The bulk of the task is identical to earlier versions (see the notes there).

Figure 10–3e Continued

```
              next[entry].pattern&=0x3;
              solenoidupdate();
           }
           else if (scan==0x1000); /*"c" key*/
           else if (scan==0x2000); /*"d" key*/
           else if (scan==0x4000){ /*"e" key*/
              if (entry) entry-;
              newnumbflg=0;
              row2update();
           }
        }
        if (scan==0x8000){ /*"f" key*/
           goflg=~goflg;
        }
     }
   }
   void main(void){
     cmx_init();
     pcmd=0x80; /*initialize solenoid I/O*/
     nowtime=newsc=goflg=0;
     for(i=0;i<20;i++){
        next[i].abstime=next[i].pattern=0;
     }
     next[0].pattern=0xff;
     cxrscre(4,&LCD); /*resource group for LCD use*/
     cxtcre(5,&slotdsp,6,display,20);
13   cxtcre(5,&slotnum,6,numdisplay,20);
     cxtcre(10,&slotsol,0,solenoid,20);
     cxtcre(5,&slotks,0,keyscan,20);
     cxtcre(5,&slotki,2,keyinterp,20);
     cxttrig(slotdsp);
     cxttrig(slotnum);
     cxttrig(slotsol);
     cxttrig(slotks);
     cxttrig(slotki);
     cmx_go();
   }
```

---

[13]The first number is the priority for the task, the third number is the number of fixed size message buffers to be reserved for incoming messages, and the final number is the amount of stack space to reserve. The numbers are somewhat arbitrary at this point.

Figure 10–3e Continued

```
#include <bbdefs.h>
#define lcddat XBYTE[0xc000]
#define lcdcmd XBYTE[0x8000]
#define pcmd XBYTE[0x2003]
#define PORTA XBYTE[0x2000]
```

1
```
BB_TCB_TYPE kin_tcb; uchar xdata kinxsp[256];
uchar xdata kincont[64]; uchar data kinisp[20];
uchar kinmsgbuf[2];
BB_TCB_TYPE key_tcb; uchar xdata keyxsp[256];
uchar xdata keycont[64]; uchar data keyisp[20];
uchar keymsgbuf[2];
BB_TCB_TYPE dsp_tcb; uchar xdata dspxsp[256];
uchar xdata dspcont[64]; uchar data dspisp[20];
uchar dspmsgbuf[20];
BB_TCB_TYPE num_tcb; uchar xdata numxsp[256];
uchar xdata numcont[64]; uchar data numisp[20];
uchar nummsgbuf[2];
BB_TCB_TYPE sol_tcb; uchar xdata solxsp[256];
uchar xdata solcont[64]; uchar data solisp[20];
uchar solmsgbuf[0];
BB_TCB_TYPE ini_tcb; uchar xdata inixsp[256];
uchar xdata inicont[64]; uchar data iniisp[20];
uchar inimsgbuf[0];

uchar LCD,events,status; uint t;

void solenoid(void){ /*solenoid drive*/
   uchar i;
   forever{
      for (i=0;next[i]!= 0xff;i++){
         if (goflg && next[i]!=0xff){ /*freeze if stopped*/
            bb_wait(next[i+1].pattern-next[i].pattern);
            PORTA=next[i++].pattern;
         }
         else {
            bb_wait(100);
            i--;
         }
      }
   }
}
void keyscan (void){
   uchar code scanpat[]={0x10,0x20,0x40,0x80};
   static uint oldsc; uint newsc; uchar j;
   forever{
```

[1]These variables are all part of the space used by the operating system. The sizes are strictly a guess at this point—probably they are much larger than necessary. The user must set out all the space for reference by the task creation system calls. It is a high level of user control, but also a bit of a pain!

Figure 10–3f Solenoid Sequencer II—Byte-BOS

```
        for(j=0,newsc=0;j<4;j++){
          P1=scanpat[j];
          newsc=(newsc<<4) | (P1 & 0xf);
          bb_wait(10); /*10 mSec interval*/
        }
2       if (newsc!=oldsc) bb_put_message(&kin_tcb,&newsc,2);
        oldsc=newsc;
      }
  }

  void display (void){
      struct {uchar cursor;uchar ch[19];}msg;
3     bb_enable_message();
      bb_wait(15); /*set up LCD modes*/
      lcdcmd = 0x30; for(t=0;t<20;t++); ;
      lcdcmd = 0x30; for(t=0;t<20;t++); ;
      lcdcmd = 0x30; for(t=0;t<20;t++); ;
      lcdcmd = 0x38; for(t=0;t<20;t++); ; /*1/16 duty?*/
      lcdcmd = 0x0e; for(t=0;t<20;t++); ; /* dsp on*/
      lcdcmd = 0x01; for(t=0;t<20;t++); ;
      lcdcmd = 0x06; for(t=0;t<20;t++); ;
      forever{
        bb_sleep_for_message(msg); /*wait for display msg*/
4       bb_capture_resource(&LCD);
        lcdcmd = 0x80 | msg.cursor; /*start location*/
        for(t=0;t<20;t++); ;
        for (i=0;i<19 && msg.ch[i]){
          lcddat = msg.ch[i];
          for(t=0;t<20;t++); ;
        }
        bb_release_resource(&LCD);
      }
  }
  void numdisplay(void){
      uint number,m; bit ldgzero;
      bb_enable_message();
```

[2]The message is "put" to another task, here the keyinterp task. This one holds only the two bytes that are newscan—the length of the message is the third parameter. An alternate form of the command (put_message_nb) will try to send the message and, if not possible due to lack of free buffers or another task in the process of sending a message to the task, will go on with a status indication. With this operating system a complete rescheduling of tasks is instituted with a message sending, so if the display or keyinterp tasks were created at higher priority, the problem of some other systems is avoided. As is, with equal priorities I'm not sure which task resumes—does the round-robin mode go on to the next task of the same priority or does it continue with the current one?

[3]Tasks will only receive messages if they have been enabled for it. Error messages return to the sender if the receiver is not enabled.

[4]This is an instance of resource management. Here the LCD is the resource to be captured so the other tasks (specifically numdisplay) do not write at the wrong cursor position. It guarantees (only by mutual agreement among programmers) that two tasks do not accidentally mix access to the LCD and mess up timing or cursor addresses.

Figure 10–3f   Continued

```
    forever{
      bb_sleep_for_mesage(&number); /*wait for msg*/
      bb_capture_resource(&LCD);
      m = 10000;
      ldgzero = 0;
      do{
         if ((((number/m)>0) || (m == 1))
            || ((m == 100) && (dpt == 1)))
            ldgzero = 1;
         if (ldgzero == 1)
            lcddat = number/m+'0';
         else lcddat = ' ';
         for(t=0;t<20;t++); ;
         number %= m;
         if (m == 100) {
            lcddat = '.';
            for(t=0;t<20;t++); ;
         }
      }
      while((m /= 10)0);
      bb_release_resource(&LCD);
   }
}

/*following 2 functions are not tasks—go with keyinterp*/

void solenoidupdate(void){
   if (next[entry].pattern & 1)
      bb_put_message(&dsp_tcb,&(92," on"));
   else bb_put_message(&dsp_tcb,&(92,"off"));
   if (next[entry].pattern & 2)
      bb_put_message(&dsp_tcb,&(81," on"));
   else bb_put_message(&dsp_tcb,&(81,"off"),5);
}

void row2update(void){
   struct {uchar loc,uint numb} msg;
   msg.loc=65; msg.numb=entry;
   bb_put_message(&num_tcb,&msg);
   numdsp(71,next[entry].abstime);
   msg.loc=71; msg.numb=next[entry].abstime;
   bb_put_message(&num_tcb,&msg);
   solenoidupdate();
}

void keyinterp (void){
   uchar x; uint scan; bit newnumbflg;
   bb_enable_message();
```

Figure 10–3f   Continued

```
        bb_put_message(&dsp_tcb,&(0,"entry"));
        bb_put_message(&dsp_tcb,&(8,"time"));
        bb_put_message(&dsp_tcb,&(15,"solenoid 1"));
        bb_put_message(&dsp_tcb,&(26,"solenoid 2"));
        forever{
5          bb_sleep_for_message(&scan); /*wait forever*/
            /*new key condition to process*/
          if (goflg==0){       /*update only if stopped*/
            if (scan==0x0000);/*0->key released*/
            else if (scanx0200){ /*0-9 key*/
                if (newnumbflg==0) {
                    next[entry].abstime=0;
                    newnumbflg=1;
                }
                for(x=0;scan & 1==0;scan>>=1)x++;
                next[entry].abstime=10*next[entry].abstime+x;
                sendnumb(71,next[entry].abstime);
            }
            else if (scan==0x0400){ /*"a" key*/
                if (entry<18) entry++;
                newnumbflg=0;
                row2update();
            }
            else if (scan==0x0800){ /*"b" key*/
                newnumbflg=0;
                next[entry].pattern&=0x3;
                solenoidupdate();
            }
            else if (scan==0x1000); /*"c" key*/
            else if (scan==0x2000); /*"d" key*/
            else if (scan==0x4000){ /*"e" key*/
                if (entry) entry-;
                newnumbflg=0;
                row2update();
            }
          }
          if (scan==0x8000){ /*"f" key*/
            goflg=~goflg;
          }
        }
    }

void main(void){
```

⁵The **sleep for** message has no timeout function. The alternative is the **wait_for_mes**sage which can timeout. The rest of the **keyinterp** is the same as virtually all the previous versions of the example.

Figure 10–3f   Continued

```
6      bb_init_bos();
       pcmd=0x80; /*initialize solenoid I/O*/
       bb_create_task(&dsp_tcb,&dspcont,&dspisp,&dspxsp,
                                        &dspmsgbuf,20,5);
       bb_create_task(&kin_tcb,&kincont,&kinisp,&kinxsp,
                                        &kinmsgbuf,0,5);
       bb_create_task(&key_tcb,&keycont,&keyisp,&keyxsp,
                                        &keymsgbuf,0,5);
       bb_create_task(&num_tcb,&numcont,&numisp,&numxsp,
                                        &nummsgbuf,20,5);
       bb_create_task(&sol_tcb,&solcont,&solisp,&solxsp,
                                        &solmsgbuf,0,5);
       bb_run_task(&dsp_tcb,display)
       bb_run_task(&kin_tcb,keyinterp)
       bb_run_task(&key_tcb,keyscan)
       bb_run_task(&num_tcb,numdisplay)
       bb_run_task(&sol_tcb,solenoid)
       bb_start_bos();
   }
```

---

[6]The start-up process is similar to earlier Byte-BOS examples— just more of it.

Figure 10–3f   Continued

```
1    #include<reg51.h>
     #include<absacc.h>
     #include<string.h>
     #include<rtxcapi.h>
     #include<ctask.h>
     #include<cclock.h>
     #include<csema.h>
     #include<cmbx.h>
     #define lcddat XBYTE[0xc000]
     #define lcdcmd XBYTE[0x8000]
     #define pcmd XBYTE[0x2003]
     #define PORTA XBYTE[0x2000]

     uchar LCD,events,status; uint t;

     void solenoid(void){ /*solenoid drive*/
        uchar i;
        forever{
```

[1]The mailboxes and semaphores are defined with a system generation utility which makes these header files to be added at the top. The details are omitted here for space savings, but the process still requires the user to define all the same things as in the other configurable RTOS.

Figure 10–3g   Solenoid Sequencer II—RTXC

```
    for (i=0;next[i]!= 0xff;i++){
        if {goflg && next[i]!=0xff { /*freeze if stopped*/
            KS_delay(0,next[i+1].pattern-next[i].pattern);
            PORTA=next[i++].pattern;
        }
        else {
            KS_delay(0,100);
            i--;
        }.
    }
}
}


void keyscan (void) {
    struct{RTXCMSG msghdr;uint newsc}msg;
    uchar code scanpat[]={0x10,0x20,0x40,0x80};
    static uint oldsc; uint newsc; uchar j;
    forever{
        for(j=0,newsc=0;j<4;j++){
            P1=scanpat[j];
            newsc=(newsc<<4) | (P1 & 0xf);
            KS_delay(0,10); /*10 mSec interval*/
        }
        if (newsc!=oldsc) KS_send(KINTERPMBX,&msg.msghdr,
                                (PRIORITY)1,nullsema);
        oldsc=newsc;
    }
{


void display (void){
    struct {RTXCMSG msghdr;uchar cursor;uchar ch[19];}msg;
    KS_delay(0,15); /*set up LCD modes*/
    lcdcmd = 0x30; for(t=0;t<20;t++);
    lcdcmd = 0x30; for(t=0;t<20;t++);
    lcdcmd = 0x30; for(t=0;t<20;t++);
    lcdcmd = 0x38; for(t=0;t<20;t++); /*1/16 duty?*/
    lcdcmd = 0x0e; for(t=0;t<20;t++); /*dsp on*/
    lcdcmd = 0x01; for(t=0;t<20;t++);
    lcdcmd = 0x06; for(t=0;t<20;t++);
    forever{
```

[2]This sending of a message involves no waiting for any acknowledgement. Presumably several messages can pile up in the k e y i n t e r p mailbox if things are busy, but in any case the scan does not need to wait for a reply.

[3]This is the normal initialization of the LCD module. Since nothing else needs to happen at the start, the software loop is used to get a short delay without the complexity of the K S_d e l a y system call.

Figure 10–3g   Continued

```
4        msg=KS_receivew(DSPMBX,0); /*wait for display msg*/
5        KS_lockw(LCD); /*wait for LCD free*/
         lcdcmd = 0x80 | msg.cursor; /*start location*/
         for(t=0;t<20;t++); ;
         for (i=0;i<19 && msg.ch[i]){
            lcddat = msg.ch[i];
            for (t=0;t<20;t++);
         }
6        KS_unlock(LCD);
7        KS_ack(msg);
      }
   }

   void numdisplay(void){
      struct{RTXCMSG msghdr;uint number}msg;
      uint number,m; bit ldgzero;
      forever{
         msg=KS_receivew(NUMMBX,0); /*wait for msg*/
         number=msg.number;
         KS_lock(LCD);
         m = 10000;
         ldgzero = 0;
         do{
            if ((((number/m)>0) || (m == 1))
                || ((m == 100) && (dpt == 1)))
               ldgzero = 1;
            if (ldgzero == 1)
               lcddat = number/m+'0';
            else lcddat = ' ';
            for(t=0;t<20;t++); ;
            number %= m;
            if (m == 100) {
               lcddat = '.';
               for(t=0;t<20;t++); ;
            }
         }
      }
```

[4]Since the display task has nothing else to do until a message arrives, the indefinite wait for a message is used. The zero specifies that messages from *any* task are acceptable. An alternative feature is to wait for a message from a particular task even though messages from other tasks may be in the mailbox. This is obviously more complex than the first few operating systems and will only happen with additional code from the operating system. There are several levels of operating system; "basic, advanced, and extended," which include more and more complex calls for increasing prices. The issue of appropriate complexity for an 8051 is discussed in Chap. 8.

[5]This resource management call seeks ownership of the LCD and waits until it is available.

[6]This returns the resource to an available condition.

[7]This acknowledges the message to the sending task. By using the s e n d w call, each task using the display task becomes blocked upon sending the message and does not resume until the reply is received. This is a somewhat more sophisticated approach, but could be emulated in other operating systems with programming which involved semaphores or waits for reply messages.

Figure 10–3g Continued

```
            while((m /= 10)>0);
            KS_unlock(LCD);
            KS_ack(msg);
        }
    }

    /*following 2 functions are not tasks—go with keyinterp*/

    void solenoidupdate(void) {
        struct{RTXCMSG msghdr;uchar pos;uchar dat[101]}msg;
8       if (next[entry].pattern & 1) strcpy(&msg.dat,"on");
        else strhcpy(&msg.dat,"off");
        msg.pos=92;
9       KS_sendw(DSPMBX,msg.msghdr, (PRIORITY)1,updsema);
        if (next[entry].pattern & 2) strcpy(&msg.dat,"on");
        else strcpy(&msg.dat,"off");
        msg.pos=81;
        KS_sendw(DSPMBX,msg.msghdr, (PRIORITY)1,updsema);
    }

    void row2update(void){
        struct {RTXCMSG msghdr;uchar pos,uint numb} msg;
        msg.pos=65; msg.numb=entry;
        KS_sendw(NDSPMBX,msg.msghdr, (PRIORITY)1,r2sema);
        msg.pos=71, msg.numb=next[entry].abstime;
        KS_sendw(NDSPMBX,msg.msghdr, (PRIORITY)1,r2sema);
        solenoidupdate();
    }

    void keyinterp (void){
        struct(RTXCMSG msghdr;uchar pos;uchar dat[10]}dmsg;
        struct{RTXCMSG msghdr;uint scan;}smsg;
10      struct {RTXCMSG msghdr;uchar pos,uint numb}nmsg;
        uchar x; bit newnumbflg;
        dmsg.pos=0; strcpy(&dmsg.dat,"entry");
        KS_sendw(DSPMBX,dmsg.msghdr, (PRIORITY)1,kinterpsema);
        dmsg.pos=8; strcpy(&dmsg.dat,"time");
        KS_sendw(DSPMBX,dmsg.msghdr, (PRIORITY)1,kinterpsema);
        dmsg.pos=15; strcpy(&dmsg.dat,"solenoid 1");
        KS_sendw(DSPMBX,dmsg,msghdr, (PRIORITY)1,kinterpsema);
        dmsg.pos=26; strcpy(&dmsg.dat,"solenoid 2");
        KS_sendw(DSPMBX,dmsg.msghdr, (PRIORITY)1,kinterpsema);
        forever{
```

[8]The strcpy library function is used here. The message contents are actually moved so the string must be installed in the structure before sending the message.

[9]Again, this sends the message to the display task and will wait until a reply (semaphore) indicates that the display task is done.

[10]There are three different message structures because there can be messages going to two sources as well one coming in from the keyscan task.

Figure 10–3g Continued

```
11      KS_receivew(KINTERPMBX,0); /*wait forever*/
        /*new key condition to process*/
        if (goflg==0) {      /*update only if stopped*/
          if (smsg.scan==0x0000); /*0->key released*/
          else if (smsg.scan<0x0200) { /* 0-9 key*/
              if (newnumbflg==0) {
                 next[entry].abstime=0;
                 newnumbflg=1;
              }
              for (x=0;smsg.scan & 1==0;smsg.scan>>=1)x++;
              next[entry].abstime=10*next[entry].abstime+x;
              nmsg.pos=71; nmsg.numb=next[entry].abstime);
              KS_sendw(NDSPMBX,msg.msghdr,(PRIORITY)1,kinterpsema);
          }
          else if (smsg.scan==0x0400) { /*"a" key*/
              if (entry<18) entry++;
              newnumbflg=0;
              row2update();
          }
          else if (smsg.scan==0x0800){ /*"b" key*/
              newnumbflg=0;
              next[entry].pattern&=0x3;
              solenoidupdate();
          }
          else if (smsg.scan==0x1000); /*"c" key*/
          else if (smsg.scan==0x2000); /*"d" key*/
          else if (smsg.scan==0x4000){ /*"e" key*/
              if (entry) entry-;
              newnumbflg=0;
              row2update();
          }
        }
        if (smsg.scan==0x8000) { /*"f" key*/
           goflg=~goflg;
        }
      }
   }
}
```

---

[11]This is an endless wait for any message. Since messages are only expected from k e y s c a n, there should be no probelm. The k e y s c a n does not expect an acknowledgement, so there is no a c k call in this task.

Figure 10–3g   Continued

REVIEW AND BEYOND

1. What is the difference between a signal and a message? When can they be used interchangeably?
2. When would a shared variable be preferable to messages or signaling? What are the advantages of the latter?
3. How could you produce a resource reserving system if it is not supported by your operating system (or if you aren't using an operating system)?
4. What is the difference between a system that actually passes the message *contents* and one that passes a message *pointer*? What cautions are necessary with the latter? What costs are associated with the former? How can a system that passes the actual contents be converted to a pointer-passing system?
5. If the sending of a message does not immediately wake the receiving task, what can the programmer do to force such an event? (Pick a specific system, if you have sufficient information.) In what situations is it important for the receiving task to signal the sending task that the message has been processed?
6. In what sort of applications could you envision a need for a dynamically allocated memory pool shared among tasks?

Chapter 11

Interrupt, Priority, and Context

INTERRUPT EVENTS

Having discussed timing and communication, we finally discuss what is most hardware related—the interrupt event. The basic hardware information on 8051 interrupts is covered in Chapter 7. The non-RTOS examples of the last two chapters have already had to use one interrupt for timing. Even with some RTOS, you will not be insulated from the interrupt initialization and the interrupt details may be up to you!

With RTOS, specific 8051 interrupts and timers may be unavailable for task use due to prior claim by the operating system. For example, when running most RTOS, the use of ET0 is forbidden because timer0 interrupt is the key to all the timing functions of the operating system itself. If you are using BITBUS distributed nodes, which work only with the 8044 chip, ES is unavailable because it is the means of exchanging messages between the processor and the serial interface unit (SIU). With the basic 8051 members, that may leave as few as three interrupts that can be used by the user's tasks (INT0, INT1, and T1). Some of the RTOS can be configured for T1 or can even be configured with *no* timing, but a timed interrupt is an essential for any time slice or timeout-capable system.

CONTEXT SWITCHING

When a particular task is running, it may be that something more urgent will need to have attention of the processor. In a *preemptive* system, rather than waiting for the first task to finish, the processor will turn at once to handle the new task and only resume the earlier task when finished. If the program flow were instantly transferred to the new task, the new task's use of the various registers could destroy any intermediate information being used by the first task. When the first task resumed, the registers might cause incorrect things to happen.

The answer to the problem is a set of instructions executed every time a change of task occurs. This is called **context switching**. In Chapter 7, the idea of register banks was mentioned. This is the mechanism inherent in the 8051 hardware design for context switching. Whenever an interrupt task is run, it can be run with a different register bank. Thus, by changing one or two bits in the P S W register, the 8 bytes (for r 0–r 7) in the one task are preserved and a different set of 8 bytes is used in the new task. Usually it is advisable to push the value of the accumulator and b register at the same time. Essentially this context switch involves two pushes and a single bit-changing instruction. At the worst, you might also save the d p h and d p l registers. That set of pushes is fairly fast, which was the intent of the hardware designers from the start.

The drawback is that only four register banks exist in the 8051 and they only hold 8 bytes each. In the original 8051 design, the intent was to have essentially three priorities. The main program could be preempted by two priorities of interrupts. The higher the priority the more interrupts were masked off, so there could be only two levels of preemption. In that way there are only three register banks needed for context switching. (The fourth was related to a special third-priority scheme which is described in the 8051 user's manual.) The registerbank count means that at most four different priority levels are possible because only three tasks could in turn be preempted by other tasks. However, many tasks of *equal* priority are possible, in a round-robin scheme, because they would not preempt each other. This two-tiered approach is the one used in DCX.

RTXtiny is not inherently preemptive so there is no context switching at all—when tasks change, the previous task has intentionally made a system call and is not in the middle of anything. This is called "cooperative multitasking" or "non-time- slice, non-preemptive multitasking."

DCE (and probably RTX51) goes one step further. It uses only two register banks—one for the running task and one for the operating system—but it had more than four priorities. To get the context switch, software copies the registers to off-chip RAM. The result is the possibility of five (or more) real priorities at one time without the limit of three available register banks. The cost is the time to copy the eight registers to off-chip RAM. The system requires the user tasks to avoid overlaying variables and the system maintains separate nonoverlapping stack pieces on-chip. The tasks must be either linked together after compiling so they are located nonoverlapping or else the user must carefully force separate locations for variables.

But, where many priorities and even hundreds or thousands of tasks are theoretically possible, some of the other operating systems have chosen to "clean out" the bulk of the on-chip memory so the many tasks can have a realistic amount of on-chip resources—particularly stack, which is the particular love of reentrant C functions. The specific details of the individual systems differ and are not clearly spelled out in some cases. If the system uses *overlapping* on-chip stack, then probably the entire stack for the task is copied to off-chip RAM. If *other* on-chip variables are to be overlaid, they too are moved to a safe place off-chip. In such cases each task can operate as though virtually all the on-chip resources are at its disposal. The operating system may either blindly copy *everything* or else it cleverly saves only the parts being used by the present task (either by information derived from the compiler/linker files or else by specific inputs from the programmer). Tasks for such systems need not be linked together if the on-chip space can effectively overlap, but the issue of conflict over off-chip RAM may be serious. For the 8051 family which doesn't have relocatable final code, *that* space can never overlap, so it is probably necessary to force locating of code and variables to different parts of off-chip RAM for different tasks as they are individually located and linked. I know of no linker that will locate multiple tasks nonoverlapping in off-chip RAM and code space while overlaying *all* on-chip resources.

The drawback of such an approach is *speed* of context switching. It will obviously take considerably longer to move 100 bytes to off-chip RAM (with the `movx` and `inc dptr` commands). As a designer you will have to choose whether you need lots of tasks and priorities with lots of variable space on-chip, or whether you need "lean and mean" operation having quick task switching. The answer will probably vary depending on the application and the size of the tasks. My own view is that it is inappropriate to use the 8051 for huge applications and that most reasonable applications can run nicely with only a few priorities, but that is admittedly a personal bias. The day may be coming with the 8051 "relatives" where the large applications can fit nicely on such chips for a cost savings over more advanced processors.

A further word on other processors is in order. Most computer systems have one address space for RAM and ROM. The space holds the program code, the variables, and the stack. Because C is a very stack-oriented language on most machines, the context can be largely held on the stack with perhaps a few other registers. Then context switching can involve swapping stack pointers to separate stack areas maintained for each task. If the stacks are large enough, the various tasks can be **reentrant** so that the *same* code can be the code for several tasks. Reentrancy is a difficult thing on the 8051 because it requires everything to be kept on a stack and the on-chip stack is quite small.

BUILT-IN INTERRUPT EVENTS

From an RTOS-user standpoint, to DCX and DCE the interrupts are simply other events, with the details handled in the operating system. Hardware interrupts are permanently assigned to specific tasks by the ITD (Initial Task Descriptor) file. Tasks created later can take over the specific interrupt because it "belongs" to the last-encountered task claiming it. For RTX51 (not RTXtiny) the same mechanism is in place but the interrupt is assigned or reassigned to a specific task by `rtx_attach_interrupt`.

INTERRUPT HANDLERS

For the other RTOS, due to the context switching concerns, interrupts are a major issue. As described in Chapter 8, from the perspective of the processor, the operating system is just some code which happens to be running. No special provisions are made. If an interrupt is unmasked and takes place, the processor pushes the return address (program counter) on the stack, masks all interrupts of equal or lower priority, and hands control to the interrupt task at the vector address. It is formidable to put such raw power in your hands, because you can trash stacks, delay time slices or timeouts, and generally foul up the system! However, you can get something done quite quickly without all the hassle of context switching. As a result, *all* the systems admit you can write *interrupt handlers*. As long as you don't unmask anything and don't use the operating system calls within the handler, all is well; you have essentially preempted the operating system. If you take too long in your handler, of course, the regular system tick will be preempted also and timeouts and time intervals will get stretched, but I assume your interrupts will be carefully written to be fast and short—if you need to do lots of processing, set a flag to be picked up in the main program or in a task.

INTERRUPTS USING THE OPERATING SYSTEM

Signaling the low-priority tasks to carry out the work leads to the next use of interrupts. If, from the interrupt handler, you want to interact with the RTOS, some sort of context switching is necessary. Some RTOS make provision for this as well. RTX-tiny avoids the issue by only allowing the sending of a *signal* to a task from the interrupt (probably just setting a bit). RTX51 allows interrupts to send and receive messages as well as assigning interrupts directly to specific tasks. USX has an indirect call, `USXCMD_C`, which can activate *any* of the normal system calls (probably by adding the necessary context switching). CMX has `cxint_in` and `cxint_ex` that inform the RTOS to save and restore the previous context; from there all the system calls can be used in the normal manner. In Byte-BOS the interrupt service routines can post events (signaling), wake up tasks, and invoke the task scheduler; if the normal system calls are to be used, the `enter_isr` and `exit_isr` calls perform the same sort of context switching as is available in CMX. RTXC provides only an interrupt service exit routine (`ISRC`) which enables the setting of one semaphore at a time within the interrupt service routine.

    Incidentally, CMX expects the user to write the interrupt to cause the system tick rather than hiding it in other code. It is included in Fig. 11–5e as another part of the overall system.

SETTING PRIORITY

In any real-time system, it is important to establish priorities. *How* to establish them is a good question, but it can be related to, *"How long can this task wait?"* For example, take a project that includes the following tasks:

1. reading a character from a serial port, which should be done before the next character arrives (typically within 1mSec);
2. recognizing a person pushing a button (this could wait for 100mSec before you begin to sense a delay); and
3. a real-time clock (*ticking*, say, at a 10mSec rate), which needs to restart the timer and count up by one before the next tick expires.

Now to set priorities:

This clock interrupt ought to be recognized *at once*, because any delay will slow down the clock (in most RTOS this is built in). You will have to balance the urgency of the interrupt with the amount of processing time it takes. Whether writing your own or with an RTOS, it is best to keep interrupt service routines as short as possible so the system can remain sensitive to *other* interrupts; set flags (or use RTOS signaling) to request that any time-consuming and less urgent processing be done in the background tasks.

For the case above, if the timer must be software reset (say in mode 1), it would be logical to make the real-time clock a high-priority interrupt, but write a very quick service routine. The serial input could be low priority, and the button scan could be a noninterrupt background task.

Then again, for the case above, the serial reception could be made high-priority because it takes only a few microseconds, and the real-time clock could be low priority but longer running to include the button scan. Your decision will depend on the impact of occasionally missing a character or occasionally slipping the clock a few microseconds. Only the specific application can determine the answer to such questions!

If you are writing without a RTOS, because the 8051 directly supports only two levels of interrupt priority (see Fig. 7–7) along with the background non-interrupt level, the use of *different* priorities should be *limited* as much as possible. In most RTOS, task interrupts will be masked when equal or higher priority tasks are running, so the use of different priorities may become unavoidable. Tasks that run quickly and are not used again for long periods of time can be put at a high priority or they can be written as ISRs without involving the operating system and context switching at all. One candidate would be a stepper drive task, which uses only a few microseconds every few mSec, yet if it gets held up too long, causes the motor to appear to "stutter."

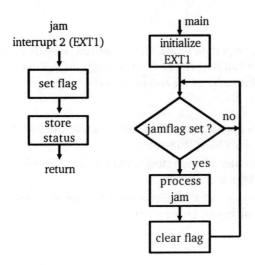

Figure 11–1a Envelope Detector

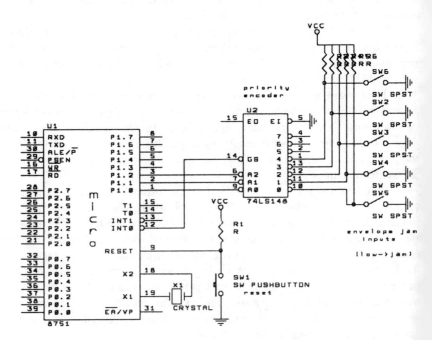

Figure 11–1b Envelope Detector

EXAMPLE: ENVELOPE DETECTOR

The envelope system of Chapter 1 could easily include hardware interrupts. Because there are numerous sensors along the path of the conveyor and stuffer systems, a single 8051 would either poll the sensors or use some sort of "wired-OR" to bring the sensor signals into a single hardware interrupt. The schematic of Fig. 11–1a shows a possible system for bringing in multiple photo-interruptor signals as interrupts using a 74148 priority encoder chip. Alternately, the simple open-collector signals from the photo-interrupters could provide a wired-OR *directly*, but then it would be impossible to separate out the source to determine *which* input caused the interrupt. There might be a suitable programmable interrupt-controller chip, but the method of pushing the vector address onto the data bus may not be compatible with the 8051's separate data and program addressing.

For this example, shown in Figures 11–2a and 11–2b in the non-RTOS versions, only the particular sensor causing the interrupt (from subsequent polling) will be sent to another task (for further processing). Each version of the example

```
#include <reg51.h>
#define uchar unsigned char
#define uint unsigned
uchar status; bit jamflg;

void jam (void) interrupt 2 using 2 {
   jamflg=1;
   status=P1;
}

void main (void){
   IP=0x04; /*set IE1 high priority*/
   IE=0x86; /*enable IE1 and IT0*/
   for(;;);
      if (jamflg) {
         switch (status){
         switch (){ /*find first jam*/
            case 0: break;/*handle jam case 1
            case 1: break; /*handle jams here*/
            case 2: break;
            case 3: break;
            default:;
         }
         jamflg=0;
      }
   }
}
```

Figure 11–2a Envelope Detector—C

```
MODULE:DO;
$INCLUDE (REG51.DCL)
   DECLARE STATUS BYTE, JAMFLG BIT;

   JAM:PROCEDURE INTERRUPT 2 USING 2;
      JAMFLG=1;
      STATUS=P1; /*ASSUME LINES TIED TO P1*/
   END JAM;

   IP=00000100B; /*SET IE1 HIGH PRIORITY*/
   IE=10000110B; /*ENABLE IE1 AND IT0*/
   WHILE 1;
      IF JAMFLG>0 THEN DO;
      DO CASE (STATUS);
         ;   /*HANDLE JAMS HERE*/
         ;
         ;
         ;
         END;
         JAMFLG=0;
      END;
   END;
END MODULE;
```

Figure 11–2b   Envelope Detector—PL/M

assumes the resources of the examples of the last two chapters, but to save space, the routines will not be repeated here.

EXAMPLE: MOTOR SPEED CONTROL SYSTEM

For a completely digital motor speed control system, it is interesting to combine a PWM drive to a DC motor with an opto-interrupter monitoring the shaft turns. Although some systems have hardware that puts out multiple pulses per turn and then the system counts pulses for a fixed time, it is also possible to determine rpm where the hardware provides only one pulse per revolution by measuring the time between pulses (one revolution). Interrupt *latency* might be of concern because the goal would be to read the time as quickly as possible after the turn is completed.

  In the specific example, where flow chart and schematic are in Figures 11–3a and 11–3b, the rpm-measuring task determines the time for a revolution by reading timer 1 after an interrupt and forwarding the difference from the last time to a control task, which does a simple up-one-or-down-one control algoithm. It might someday provide Proportional-Integral-Derivative (PID) algorithms, but such calculations are not the focus here. The control task sends a new "width" to a pulse task revised from Chapter 9. Further design issues are discussed in the RTOS version of the examples shown in Figures 11–4a and 11–4b.

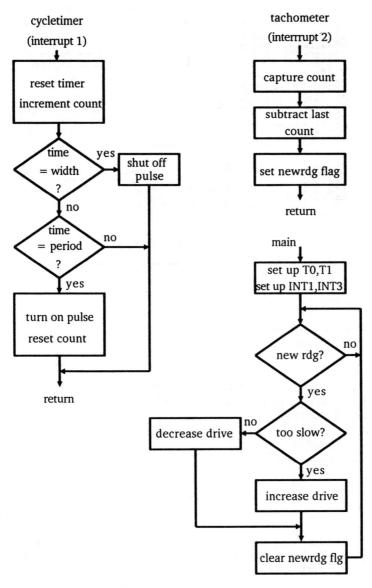

Figure 11–3a Speed Control

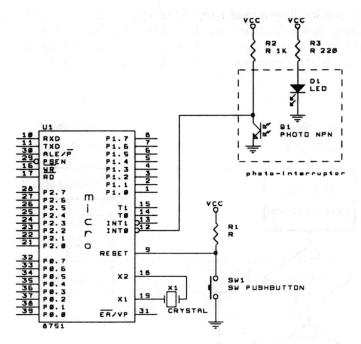

Figure 11–3b Speed Control

```
#include <reg51.h>
#define uchar unsigned char
#define uint unsigned
uchar status,percent,period;
bit newreadingflg;
uint target=500; /*arbitrary starting goal*/
```

```
1  void pulse (void) interrupt 1 using 1{
     TH0 = -833/256;/*1mSec—10MHz crystal*/
     TL0 = -833%256;
     ET0 = 1;
     if (++nowtime==percent) P1=0;
     else if (nowtime==100){
       nowtime=0;
       P1=1;
     }
   }
```

¹This interrupt comes every millisecond. It sets or clears the output to a port bit to produce a PWM signal to a DC motor. The overall period is a constant 100mSec while the port is high for whatever value is in percent.

Figure 11–4a Speed Control—C

```
2  void tachometer(void) interrupt 1 using 1 {
      union split{uint word;
                  struct{uchar hi;uchar low;}bytes}};
      uint oldcount;
      newcount.bytes.hi=TH1;
      newcount.bytes.lo=TH0;
      period=newcount.word-oldcount;
      oldcount=newcount.word;
      newreadingflg=1;
   }

   void main (void){
      IP=0x04; /*set IE1 high priority*/
      TMOD=0x11; /*T1 in endless 16-bit overflow mode*/
      TCON=0x54; /*T0,T1 running, IE1 edge triggered*/
3     TH1=0; TL1=0; /*set initial count (unnecessary)*/
      IE=0x8c; /*enable IE1 and IT1*/
      for(;;);
         if (newreadingflg) {
4            if (period < target){
                if (percent<100) pulse(++percent);
             }
             else if (percent>0) pulse(−percent);
             newreadingflg=0;
          }
       }
    }
```

---

[2]This interrupt is driven by an opto-interruptor triggered by a mark on the motor shaft. It determines the time for a revolution of the shaft by comparing the reading of timer 1 with the reading the last time. It signals the control algorithm by setting a flag bit.

[3]Timer 1 is let go in a 16-bit mode without ever reloading it. It provides a convenient 16-bit value which just folds over endlessly. You should very carefully keep that timer interrupt masked!

[4]This algorithm is quite primitive and updates at a rate determined by the motor speed. It would be a sad example for a controls course.

Figure 11–4a Continued

```
1 MODULE: DO;
  $INCLUDE (REG51.DCL)
  DECLARE(NEWCOUNT,OLDDCOUNT,TARGET)WORD;
  DECLARE PERCENT BYTE, NEWREADINGFLG BIT;

  PULSE:PROCEDURE INTERRUPT 1 USING 1;
     TH0 =HIGH(-833);/*1mSec(10MHz OSC)*/
     TL0 =LOW(-833);
     ET0 = 1;
     NOWTIME=NOWTIME+1;
     IF NOWTIME=PERCENT THEN P1=0; /*END PULSE WIDTH*/
     ELSE IF NOWTIME=100 THEN DO;
        NOWTIME=0;
        P1=1; /*START NEXT PULSE*/
     END;
  END PULSE;

  TACHOMETER:PROCEDURE INTERRUPT 2 USING 2;
     NEWCOUNT=SHL(DOUBLE(TH1),8) OR TL1;
     PERIOD=NEWCOUNT-OLDCOUNT;
     OLDCOUNT=NEWCOUNT;
     NEWREADINGFLG=1;
  END TACHOMETER;

  IP=00000100B; /*SET IE1 HIGH PRIORITY*/
  TMOD=00010001B; /*T1 IN ENDLESS 16-BIT OVERFLOW MODE*/
  TCON=01010100B; /*T0,T1 RUNNING, IE1 EDGE TRIGGERED*/
  TH1=0; TL1=0; /*SET INITIAL COUNT (UNNECESSARY)*/
  IE=10001100B; /*ENABLE IE1 AND IT1*/
  DO FOREVER;
     IF NEWREADINGFLG>0 THEN DO;
         /*PRIMITIVE CONTROL ALGORITHM!
            REPLACE FOR BETTER CONTROL*/
        IF PERIOD < TARGET THEN DO;
           IF PERCENT<100 THEN PERCENT=PERCENT+1;
        END;
        ELSE IF PERCENT>0 THEN PERCENT=PERCENT-1;
        NEWREADINGFLG=0;
     END;
  END;
END MODULE;
```

¹This is a direct translation of the C version—see the comments there.

Figure 11–4b   Speed Control—PL/M

EXAMPLE: ENVELOPE DETECTOR *WITH* RTOS

There is more difference *between* the operating system approaches than there is between them and the non-RTOS approach. The difference goes back to the interrupt handler issue—does the operating system handle the interrupts directly or do you handle it more or less *outside* the operating system? Only DCX and DCE can isolate you from writing a true interrupt function.

The examples shown in Figures 11–5a through 11–5g, assume some of the same hardware as the example of Chapter 10 is in use even though it isn't shown on the schematic. There would be conflict over the use of P1 for the priority encoder and for the keyscan, of course. Assume, if you like, that the keypad is moved over to PORTC of the 8255 which can set the upper and lower nibbles for different directions.

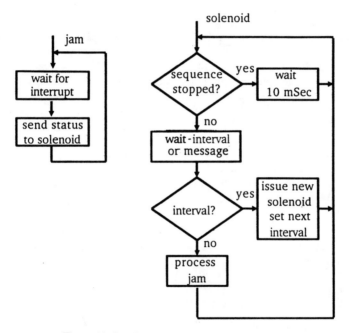

Figure 11–5a   Envelope Detector Task Flow Charts

```
1        $INCLUDE(RMX51A.MAC)
         EXTRN CODE (solenoid,initialize,jam)
         CSEG AT 0FFF0H
         ITD1:
         %ITD(ENTRY1,10,81H,101B,1,0,ITD1+1)
         ENTRY1:LJMP initialize
         CSEG AT 8000H
         ITD2:
2        %ITD(ENTRY1,10,83H,110B,2,0,ITD3)
         ENTRY2:LJMP solenoid
         ITD3:
3        %ITD(ENTRY1,10,87H,111B,3,00100b,ITD3+1)
         ENTRY7:LJMP jam
         END

         *****************************************************
         #include <dcx1.h>

3    void jam (void) {
         struct {struct header hdr;uchar status}*jmsg;
         forever{
            rqwait(0xff,interrupt);
            if (jmsg=rqallocate()){ /*buffer available*/
               jmsg->hdr.len=8;
               jmsg->hdr.nod=0;
               jmsg->hdr.flg=0;
               jmsg->hdr.cmd=0;
               jmsg->hdr.sdt=rqtaskid<<4 | taskof(0x83);
               jmsg->status=P1;
               rqsendmessage(jmsg); /*send to solenoid task*/
            }
         }
    }

         *****************************************************
         #include <dcx2.h>
         #define solenoids XBYTE[0X2000]
         #define portcmd XBYTE[0X2003]

         void solenoid(void){
```

[1]There would be other ITD entries for other tasks if this were an extension of the Chapter 1(example.

[2]The solenoid task is made higher priority so it can preempt the other (not shown) tasks tha` might be displaying or computing. Since the update of the solenoid states is quite quick, there is no good reason to allow it to wait in queue here.

[3]The jam task is made highest priority to allow immediate response to a jam. Probably it ough` to address the solenoid directly rather than pass a message, but this illustrates the mechanics o` communicating out of an interrupt function.

Figure 11–5b Envelope Detector—DCX and DCE

```
        portcmd=0x80; /*initialize 8255*/
        forever{
          for (i=0;next[i].pattern!=0xff;i++){
          if (goflg){ /*freeze count if stopped*/
            solenoids=next[i].pattern;
            smsg=rqwait(next[i+1].abstime-next[i].abstime,
                                    timeout | message);
            if (status=message){
              rqdeallocate(smsg);
              while (P1>0){ /*until jam is cleared*/
                /* sound alarm, shut off motors, etc.*/
              }
            }
          }
          else {
            rqwait(10,timeout); i--;
          }
        }
      }
```
(left margin markers: **4** aligned with the `smsg=rqwait(...)` line, **5** aligned with the `rqwait(10,timeout); i--;` line)

```
      ********************************************************

      #include <dcx1.h>

      void initialize (void){
```
(left margin marker: **6**)

```
        /*see Fig. 10-6a for details*/

        rqcreatetask(0x806a); /*jam*/
        rqdeletetask(rqtaskid); /*deletes itself*/
      }
```

[4]The solenoid task waits both for a timeout (meaning the next solenoid condition can be sent) and for a message (which means there has been a jam).

[5]This allows other tasks to run while system is stopped.

[6]It is not necessary to initialize the interrupts because the operating system handles it in response to the indications in the ITD.

Figure 11–5b   Continued

```
        #include <reg51.h>
        #include <absacc.h>
        #include <RTXtiny.h>
        #define solenoids XBYTE[0X2000]
        #define portcmd XBYTE[0X2003]
1       uchar envelopestatus;

2       void jam (void) interrupt 2 using 2 _task_ 7{
            envelopestatus=P1;
            isr_send_signal( ); /*send to solenoid task*/
          }
        }
    }

3   void solenoid(void) _task_ 2{
      forever{
        for (i=0;next[i].pattern!=0xff;i++){
        if (goflg){ /*freeze count if stopped*/
          solenoids=next[i].pattern;
4         status=os_wait(K_TMO | K_SIG,
                    next[i+1].abstime-next[i].abstime);
          if (status==K_SIG){
            while (P1>0){ /*until jam is cleared*/
              /*sound alarm, shut off motors, etc.*/
            }
          }
        }
        else {
          os_wait(K_TMO,10); i--;
        }
      }
    }

    void initialize (void) _task_ 0 {
      portcmd=0x80; /*initialize 8255*/
5     IP=0x04; /*set IE1 high priority*/
      IE=0x86; /*enable IE1 and IT0*/

      /*same as Fig. 10-6b here*/

      os_create(6); /*jam*/
      os_delete(0); /*initialize deletes itself*/
    }
```

[1]The tasks share the status byte which indicates which place the jam occurred.

[2]The interrupt task can only send a signal to other tasks. Here it signals that a jam has happened and the specifics are in the shared byte, envelopestatus.

[3]This walks through the solenoid sequence of Chapter 10 with the added wait for a signal from the jam task.

[4]The os_wait can involve waiting for either of two (or more) events—here a signal or a timeout.

[5]The interrupts must be initialized to be used.

Figure 11–5c Envelope detector—RTXtiny

```
#include  <reg51.h>
#include  <absacc.h>
#include  <usxcfg.h>
#include  <usx.h>
#define  forever  for(;;)
#define  uchar  unsigned  char
#define  uint  unsigned
#define  solmbx  4
#define  solenoids  XBYTE[0X2000]
#define  portcmd  XBYTE[0X2003]
uchar  slotini,slotsol;
uchar  xdata  free_memory[MEM_ALLOCATION];
```

```
1   void jam (void) interrupt 2 using 2 {
        uchar envelopestatus;
        envelopestatus=P1;
2       usxcmd_c(SNDMSG,solmbx,&envelopestatus,100);
    }

    void solenoid(void) {
        uchar *envelopestatus;
        forever{
            for (i=0;next[i].pattern!=0xff;i++){
            if (goflg){ /*freeze count if stopped*/
                solenoids=next[i].pattern;
                dlytsk(slotsol,
                        next[i+1].abstime-next[i].abstime);
3               if (chkmsg(solmbx)){
                    envelopestatus=rcvmsg(solmbx,0);
                    while (P1>0){ /*until jam is cleared*/
                        /*sound alarm,shut off motors, etc.*/
                    }
                }
            }
            else {
                dlytsk(slotsol,1); i--;
            }
        }
    }

    void main(void{
        portcmd=0x80; /*set up 8255*/
```

[1]Notice that jam is not an endless loop—it is an interrupt rather than a task to USX.

[2]System calls can be invoked only indirectly as shown. It probably leads to a context switch.

[3]It is not possible to produce a delay with the possibility of a message all in one call. Here the checking for a message comes only at the end of each step in the sequence.

Figure 11–5d   Envelope Detector—USX

```
    MEM_DEF *memptr;
    usxini();
    usrclk_init();
    memptr=(MEM_DEF *)free_memory;
    memptr->mem_size=
                MEM_ALLOCATION-sizeof(struct mem_def);
    memptr->task_slt=SYSTEM;
    memptr->mem_nxt=NULL;
    relmem((usxmem_t *)(++memptr));
    IP=0x04; /*set IE1 high priority*/
    IE=0x86; /*enable IE1 and IT0*/

    /*other task initialization here*/

    slotjam=runtsk(10,&jam,20);
    solslot=runtsk(200,&initialize,20);
    begusx();
}
```

Figure 11–5d Continued

```
#include <reg51.h>
#include <absacc.h>
#include <cmxfunc.h>
#define forever for(;;)
#define uchar unsigned char
#define uint unsigned
#define solenoids XBYTE[0X2000]
#define portcmd XBYTE[0X2003]

uchar slotsol,slotjam;
uchar status;

void jam (void) interrupt 2 {
    uchar envelopestatus;
    cxint_in();
    envelopestatus=P1;
    cxfgsig(slotsol,envelopestatus,0xff);
    cxint_ex();
```

1 void jam (void) interrupt 2 {

[1]Jam is an interrupt rather than a task. It is not an endless loop and relinquishes control only when the function exits. The cxint_in() call saves the previously running tasks context. In CMX it is dangerous to do anything in the interrupt before saving the context, and some C compilers will automatically save a stack frame which messes things up before you get to the cxint_in. It is recommended that the interrupt start out in assembly to avoid that, but Franklin C should be no problem. It might be that using a separate register bank would help, but CMX prefers to stay in the default bank.

Figure 11–5e Envelope Detector—CMX

```
    }

void solenoid(void){ /*solenoid drive*/
    uchar envelopestatus,i,status;
    forever{
        for (i=0;next[i]!= 0xff;i++){
            if (goflg && next[0]!=0xff){ /*freeze if stopped*/
```
2
```
                status=cxfgwatm(0,0xff,0xff,
                            next[i+1].pattern-next[i].pattern);
```
3
```
                if (status) solenoids=next[i++].pattern;
                else {
                    while (P1>0){ /*until jam is cleared*/
                    /* ought to sound alarm here*/
                    }
                    cxfgsig(slotsol,0,0xff); /*clear flags*/
                }
                else {
                    cxtwatm(100);
                    i--;
                }
            }
        }
    }
}
```
4
```
void timer0_int (void){
    cxint_in();
    TL0=10000%256;
    TH0=10000/256;
    cmx_tic();
    cxint_ex();
}

void main(void){
    portcmd=0x80; /*set up PORTA*/
    cmx_init();
    IP=0x04; /*set IE1 high priority*/
    IE=0x86; /*enable IE1 and IT0*/

    /*other task initialization here*/

    cxtcre(5,&slotjam,0,timetask,20);
    cxtcre(5,&slotsol,0,timetask,20);
    cmx_go();
}
```

---

[2]The wait for a flag gets double mileage by providing the interval for the solenoid as well. This way the system remains sensitive to jam messages all the time it is waiting for the next solenoid state change time.

[3]Status is zero if a flag arrived.

[4]This is the interrupt that drives the system tick. You must write it if you want any timeouts to happen—something must call cmx_tic. Interrupt latency and the actual activity produced by the C compilers is a concern. It is suggested you write this in assembly or at least compile once with the code option to be sure nothing is messing up the context.

Figure 11–5e   Continued

```
     #include <reg51.h>
     #include <absacc.h>
     #include <bbdefs.h>
     #define solenoids XBYTE[0X2000]
     #define portcmd XBYTE[0X2003]
     BB_TCB_TYPE jam_tcb; uchar xdata jamxsp[256];
     uchar xdata jamcont[64]; uchar data jamisp[20];
     uchar jammsgbuf[0];
     BB_TCB_TYPE sol_tcb; uchar xdata solxsp[256];
     uchar xdata solcont[64]; uchar data solisp[20];
     uchar solmsgbuf[0];

1    void jam (void) interrupt 2 using 2 {
         uchar envelopestatus;
         bb_enter_isr();
         envelopestatus=P1;
         bb_put_message_nb(&sol_tcb,&envelopestatus,1);
         bb_exit_isr();
     }

     void solenoid(void){ /*solenoid drive*/
         uchar envelopestatus,i,status,len;
         bb_enable_message();
         forever{
             for (i=0;next[i]!= 0xff;i++){
                 if (goflg && next[0]!=0xff){ /*freeze if stopped*/
2                    bb_wait_for_message(&envelopestatus,&len,
                                 next[i+1].pattern-next[i].pattern);
3                    if (status) solenoids=next[i++].pattern;
                     else {
                         /*process a jam here*/
                     }
                 }
                 else {
                     bb_delay(100); i--;
                 }
             }
         }
     }
```

[1]The interrupt is *not* a task in the normal sense. It actually interrupts the operating system and runs "on its own." The `enter_isr` call saves the context so the interrupt can proceed to run as a normal task.

[2]This high-level call combines a timeout and a wait for a message. The timeout handles the normal sequencer delay between states.

[3]Status is returned zero if a message is received.

Figure 11–5f   Envelope Detector—Byte-BOS

```
void main(void){
   portcmd=0x80; /*set up 8255*/
   bb_init_bos();
   IP=0x04; /*set IE1 high priority*/
   IE=0x86; /*enable IE1 and IT0*/
   bb_create_task(&jam_tcb,&jamcont,&jamisp,
                          &jamxsp,&jammsgbuf,4,5);
   bb_create_task(&sol_tcb,&solcont,&solisp,
                          &solxsp,&solmsgbuf,4,5);
   bb_run_task(&sol_tcb,solenoid)
   bb_run_task(&jam_tcb,jam)
   bb_start_bos();
}
```

Figure 11–5f   Continued

EXAMPLE: SPEED CONTROL *WITH* RTOS

In the non-RTOS versions, some critical design decisions were glossed over. With the RTOS examples (Figures 11–6a through 11–6g), if the system tick is frequent enough, the inherent timing of the RTOS may suffice. Is a 1mSec resolution of time sufficient resolution if you count the time for one revolution? At 6000 rpm that is 10 mSec per revolution, or an accuracy of 10 percent! At 600 rpm, however, 1 percent might be quite acceptable as part of a control loop.

If you prefer, you can count pulses for a fixed time. How often do you need a new reading? If you count for 1 second, you resolve 1 part in 100 at 6000 rpm— back to 1 percent!

So you can redesign the hardware to put out *multiple* pulses per revolution and count for a fixed time. That suits the typical control algorithms which prefer regular time intervals. If it produces 100 pulses per revolution, then the 600 rpm gives 1000 counts (0.1% resolution) and the 6000 rpm resolves 0.01%. Now you might count for only 100mSec to get update information more quickly while still having a reasonable resolution.

But suppose the cost of the 100 pulse shaft encoder is too much— what then? A single pulse per revolution system can still work. Even at 600 rpm there are 10 turns to time each second—probably a very adequate update frequency. If the *other* timer is used, very good resolution is possible by doing a capture of the timer value every time the pulse arrives. Now a resolution of 1uSec is obtained, which is 0.01 percent at 6000 rpm. That is clearly enough. This approach is used for the examples because it provides a good mix of handlers and built-in interrupt events where available.

In the same vein, the larger systems do not dream of a 1mSec tick. If the

```
    #include <reg51.h>
    #include <absacc.h>
    #include <rtxcapi.h>
    #include <csema.h>
    uchar envelopestatus;
    SEMA semalist[]={solsema,jamsema,0};

    frame jam (void) interrupt 2 using 2 {
      envelopestatus=P1;
1     return isrc(frame,jamsema);
    }

    void solenoid(void){ /*solenoid drive*/
      uchar i,status;
      forever{
        for (i=0;next[i]!= 0xff;i++){
          if (goflg && next[0]!=0xff){ /*freeze if stopped*/
2           KS_start_timer(timer,
                      next[i+1].pattern-next[i].pattern,0,solsema);
3           status=KS_waitm(semalist);
            if (status==solsema) solenoids=next[i++].pattern;
            else {
              /*process a jam here*/
            }
          }
          else {
            KS_delay(0,100); i--;
          }
        }
      }
    }
```

[1]This is the only possibility within an interrupt service routine. The precise form is not clear from the data book, but the general idea is that the parameter passed is the indication of a particular semaphore to be set. The call can be made repeatedly for multiple signals, but apparently there is no context switching done by the operating system and the system cannot be reentrant.

[2]In order to keep the solenoid timers running, it is necessary to start a timer for the time until the next update because it is necessary to have a semaphore—otherwise a simple delay could suffice. This timer is started in a one-shot mode (the third parameter zero) because it will not need to repeat but will be loaded with the next time interval value when the time expires.

[3]The wait for multiple events is necessary to allow the presence of a jam to wake the task early. The wait is for any of the events in the semaphore list. The returned status value indicates which semaphore (event) happened. Like several other high-end RTOS, RTXC seems to be written with lots of specific defines which are supposed to make it generic across processors. In this case it is not clear what such terms as solsema and timer actually represent to the compiler—the "magic" is buried in the header file produced by the configuration utility. This is rather unsettling to a hardware-and-assembly-language programmer!

Figure 11–5g   Envelope Detector—RTXC

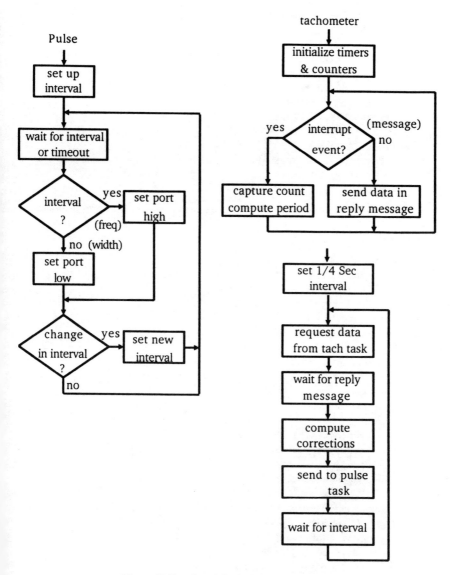

Figure 11–6a Speed Control Task Flow Charts

percent drive is to be handled in software (the "relatives" have beautiful Pulse Width Modulation [PWM] outputs which avoid the entire problem), something special will have to be done if the DC motor is not to sound like it is being hit with a hammer every pulse interval. An integrator and power amplifier could provide a varying but "smooth" voltage to the motor at whatever time constant you wish, or else the timer that is working for the tachometer can also do for the motor.

```
$INCLUDE(RMX51A.MAC)
 EXTRN CODE (tachometer,control,pulse)
 CSEG AT 0FFF0H
 ITD1:
```
1
```
 %ITD(ENTRY1,10,81H,110B,3,01000B,ITD2)
 ENTRY1:LJMP tachometer
 CSEG AT 8000H
 ITD2:
 %ITD(ENTRY1,10,82H,101B,2,0,ITD3)
 ENTRY2:LJMP control
 ITD3:
 %ITD(ENTRY1,10,83H,101B,2,0,ITD3+1)
 ENTRY3:LJMP pulse
 END

 ****************************************************
```
1
```
#include <dcx2.h>

void tachometer(void) {
struct {struct header hdr;uint period}*cmsg;
union split{uint word;
            struct{uchar hi;uchar low;}bytes}};
union split newcount;
uint newestperiod,oldcount;
```
2
```
IP=0x04; /*set IE1 high priority*/
TMOD=0x11; /*T1 in endless 16-bit overflow mode*/
TCON=0x54; /*T0,T1 running, IE1 edge triggered*/
TH1=0; TL1=0; /*set initial count (unnecessary)*/
IE=0x8c; /*enable IE1 and IT1*/
forever{
  newcount.bytes.hi=TH1;
  newcount.bytes.lo=TH0;
  newestperiod=newcount.word-oldcount;
  oldcount=newcount.word;
```
3
```
do {
    cmsg=rqwaitm(0xff,message | interrupt);
    if (status==message) {
      cmsg->hdr.flg|=0x80; /*a reply*/
      cmsg->period=newestperiod;
      rqsendmessage(cmsg);
```

[1]This task must be a different priority and use a different bank because it relies on the external interrupt from the opto-sensor.

[2]Since it relies on the external interrupt and the use of T1, it is necessary to set things up once outside the forever loop. Note that this is still a *task* rather than an interrupt.

[3]The do is used because the rqwait is needed before status can be known. The wait does double duty waiting for a message or an external event. The message is a request from the control task to get new data.

Figure 11–6b   Speed Control—DCX and DCE

```
           }
         }
      while ((status&interrupt)==0);
         }
   }

***********************************************************

   #include <dcx1.h>
   uchar percent;

   void pulse(void){
      struct {struct header hdr;uchar percent}}*msg;
4     rqsetinterval(100);
      forever{
5        msg=rqwaitm(percent,message|interval);
         if (status=message){
            percent=msg->percent;
            rqdeallocate(msg);
         }
         else P1=(status)?0:1;
      }
   }

   void control (void){
      struct {struct header hdr;
                 union{uint period;uchar percent}}*mmsg;
      uint period,target;
      forever;
         rqsetinterval(250); /*4 updates per second*/
         forever{
              /*request new data from tachometer*/
            if (mmsg=rqallocate()){ /*non-zero if valid buffer*/
            mmsg->hdr.len=9;
            mmsg->hdr.nod=mmsg->hdr.flg=mmsg->hdr.cmd=0;
            mmsg->hdr.sdt=rqtaskid<<4 | taskof(0x81);
            rqsendmessage(mmsg);
            }
6           mmsg=rqwaitm(10,message);
            if (status==message){
```

[4]The later RTOS versions do not have the speed to handle 1mSec time units, but here the basic period is set to 100mSec.

[5]The wait here has triple duty! If the period expires, the pulse should be set low; if the interval arrives, a new pulse should be started; and if a message arrives, the pulse width (percent) should be changed. The later RTOS versions avoid this structure because it is more difficult to support—they use a simple timeout for both the on and the off time.

[6]If the tachometer replies within 10 mSec, a new value is computed. Otherwise the task goes down to wait for the next 250mSec interval, giving updates at a regular time interval unlike the earlier, non-RTOS example.

Figure 11–6b   Continued

```
7                period=mmsg->period;
                 if (period < target) if (percent<100) ++percent;
                 else if (percent>0) -percent;
                    /*send data to pulse task*/
                 mmsg->hdr.len=8;
                 mmsg->hdr.nod=mmsg->hdr.flg=mmsg->hdr.cmd=0;
                 mmsg->hdr.sdt=rqtaskid<<4 | taskof(0x83);
                 mmsg->percent=percent;
                 rqsendmessage(mmsg);
              }
           rqwait(0xff,interval);
        }
     }
}
```

[7]A message need not be allocated because the one coming back from the tachometer is turned around and sent to the pulse task.

Figure 11–6b   Continued

```
#include <reg51.h>
#include <RTXtiny.h>
uchar percent;
uint period,target;

void tachometer(void) interupt 3 using 2 {
   union split{uint word;
                      struct{uchar hi;uchar low;}bytes}};
   union split newcount;
   uint oldcount;
   newcount.bytes.hi=TH1;
   newcount.bytes.lo=TH0;
1  period=newcount.word-oldcount;
   isr_send_signal(); /*tell control new value ready*/
   oldcount=newcount.word;
}

void control (void) _task_ 0 {
   os_create (1); /*pulse task*/
   os_create (2); /*interval task*/
   IP=0x04; /*set IE1 high priority*/
2  TMOD=0x11; /*T1 in endless 16-bit overflow mode*/
   TCON=0x54; /*T0,T1 running, IE1 edge triggered*/
```

[1]Here the tachometer interrupt initiates the transfer of new data by signaling the control task. The control task is quite free to ignore the signal if it chooses, but it has assurance that there are new data available.

[2]The control task which is the first to come up initializes the interrupts and timer 1.

Figure 11–6c   Speed Control—RTXtiny

```
     TH1=0; TL1=0; /*set initial count (unnecessary)*/
     IE=0x8c; /*enable IE1 and IT1*/
     forever{
3        os_wait(K_TMO,250); /*update every 1/4 Sec*/
4        if(os_wait(K_SIG,0)==K_SIG){
            os_clr_signal(0); /*itself*/
            if (period < target) if (percent<100) ++percent;
            else if (percent>0) --percent;
            /*assumes percent is shared with pulse task*/
         }
     }
  }

  void pulse(void) _task_ 1{
     for(;;){
5        P1=1;
         os_wait(K_TMO,percent);
         P1=0;
         os_wait(K_TMO,100-percent);
     }
  }
```

---

[3]The regular update time is set by the 250mSec timeout. It is *not* an interval so the time could stretch out if the control calculations stretch beyond a system tick (here assumed 1mSec).

[4]The signal should have come from the tachometer that a new reading is available. Otherwise the system waits for the next 1/4Sec update time.

[5]This routine is revised from the one of Chapter 9—this is simpler and there is nothing special needing to be shown here.

Figure 11–6c   Continued

```
#include <reg51.h>
#include <usxcfg.h>
#include  <usx.h>
#define forever for(;;)
#define uchar unsigned char
#define uint unsigned
#define ready 1
#define ctrlmbx 4
uchar slotctrl,slotpulse;
uchar xdata free_memory[MEM_ALLOCATION];
uint target;
```

```
1   void tachometer(void) interupt 3 using 2 {
        union split{uint word;
                    struct{uchar hi;uchar low;}bytes}};
        union split newcount;
        uint oldcount,period;
        newcount.bytes.hi=TH1;
        newcount.bytes.lo=TH0;
        period=newcount.word-oldcount;
2       usxcmd_c(setevt,ready);
        oldcount=newcount.word;
    }

    void control (void) {
        uint period;
        IP=0x04; /*set IE1 high priority*/
3       TMOD=0x11; /*T1 in endless 16-bit overflow mode*/
        TCON=0x54; /*T0,T1 running, IE1 edge triggered*/
        TH1=0; TL1=0; /*set initial count (unnecessary)*/
        IE=0x8c; /*enable IE1 and IT1*/
        period(ctrlintvl,25);
        forever{
4           wteset(ctrlintvl,0); /*update every 1/4 Sec*/
            decevt(ctrlintvl);
5           if (chkevt(ready)){
                clrevt(ready);
```

[1]This is an interrupt rather than a task. It cannot be a forever loop. It is not clear whether a separate register bank is needed.

[2]An interrupt cannot receive an event signal, so it has to *send* a signal to the other task saying that a new reading is available.

[3]The initialization of timer 1 and the external interrupt for the tachometer can be done here or in the initialization.

[4]To keep the period between updates of the control parameter constant, an interval of 25×10 mSec is used.

[5]The ready event comes from the tachometer every time it gets a new reading. If a new reading has not arrived, the control update is not done. Note that here the event is set and cleared rather than incremented and decremented. If multiple new readings have arrived in the 250mSec, it is only the latest one that is desired.

Figure 11–6d   Speed Control—USX

```
        if (period < target){
            if (percent<100) ++percent;
        }
        else if (percent>0) --percent;
        sndmsg(pulsembx,&percent,5);
    }
  }
}

void pulse(void){
  for(;;){
6     P1=1;
      dlytsk(slotpls,1,percent);
      P1=0;
      dlytsk(slotpls,1,100-percent);
  }
}

void main(void{
  MEM_DEF *memptr;
  usxini();
  usrclk_init();
  memptr=(MEM_DEF *)free_memory;
  memptr->mem_size=
              MEM_ALLOCATION-sizeof(struct mem_def);
  memptr->task_slt=SYSTEM;
  memptr->mem_nxt=NULL;
  relmem((usxmem_t *)(++memptr));

  /*other task initialization here*/

  slotctrl=runtsk(10,&pulse,20);
  slotctrl=runtsk(5,&control,20);
  begusx();
}
```

---

[6]This is simplified from Chapter 9 because there is no real need for an interval.

Figure 11–6d   Continued

```
#include <reg51.h>
#include <cmxfunc.h>
#define forever for(;;)
#define uchar unsigned char
#define uint unsigned
uchar slotini,slotc,slotp,slotintvl;
uint target; uchar percent;

void tachometer(void) interupt 3 {
    union split{uint word;
                struct{uchar hi;uchar low;}bytes}};
    union split newcount;
    uint oldcount,period; uchar stat;
    cxint_in();
    newcount.bytes.hi=TH1;
    newcount.bytes.lo=TH0;
1   if ((cxfgread(slotc,&stat)==0)&&(stat&1)){
                        /*tell control new value*/
        cxmssend(slotc,&(period=newcount.word-oldcount));
        cxesig(slotc,0); /*clear the flag*/
    }
    oldcount=newcount.word;
    cxint_ex();
}

void control (void) {
    uint period;
    forever{
2       cxtwatm(250); /*update every 1/4 Sec*/
3       cxfgsig(slotc,1,1); /*request newest reading*/
4       if(cxmswatm(&period,10)==0){
            if (period < target){
                if (percent<100) ++percent;
            }
```

¹This version reads a flag signal indicating that the control task requests a new reading. Then it sends the latest reading as a mesage to the control task. Note again that this is not a task and has to have the context switching calls at the start and the end.

²This timeout, assuming a 10mSec system tick, gives a constant update period of 250mSec unless the computations let it slip beyond the next tick—it isn't an interval so it doesn't keep counting if the task doesn't re-enter the wait before the next tick.

³Here the flag is set (actually belonging to this task) which is read by the interrupt to determine if a new value is desired.

⁴Here is a wait for the reply (a message) from the interrupt holding the period reading. If it doesn't arrive in 100mSec, the wait times out and returns to the 250mSec wait. Clearly this can cause significant stretch in the update times for the control algorithm since the interrupt won't even see the flag until an interrupt arrives from the opto-interruptor. Perhaps it would be better to set up a *separate* task which holds the most recent reading at all times and can respond at once to a request for new data. At least, you should be getiing a taste for some of the trade-offs you face.

Figure 11–6e   Speed Control—CMX

```
          else if (percent>0)  - -percent;
      }
    }
  }

  void pulse(void);
    forever{
5     P1=1;
      cxtwatm(percent);
      p1=0;
      cxtwatm(100-percent);
    }
  }

  void main(void){
    cmx_init();
    TMOD=0x11; /*T1 in endless 16-bit overflow mode*/
    TCON=0x54; /*T0,T1 running, IE1 edge triggered*/
    IP=0x04; /*set IE1 high priority*/
    IE1=1; IT1=1;
    cxecre(0,&updatereq); /*event group for control*/
    cxtcre(10,&slotp,pulse,20);
    cxtcre(5,&slotc,1,control,20);
    cmx_go();
  }
```

[5]This is simplified from Chapter 9 to make it faster running and easier to follow. Percent is actually shared among tasks rather than passed as a message.

Figure 11–6e Continued

```
#include <reg51.h>
#include <bbdefs.h>
#define interval 0x0001
#define updatereq 0x0002
BB_TCB_TYPE tac_tcb; uchar xdata tacxsp[256];
uchar xdata taccont[64]; uchar data tacisp[20];
uchar tacmsgbuf[2];
BB_TCB_TYPE pls_tcb; uchar xdata plsxsp[256];
uchar xdata plscont[64]; uchar data plsisp[20];
uchar plsmsgbuf[2];
BB_TCB_TYPE ctl_tcb; uchar xdata ctlxsp[256];
uchar xdata ctlcont[64]; uchar data ctlisp[20];
uchar ctlmsgbuf[2];
BB_TCB_TYPE sol_tcb; uchar xdata solxsp[256];
uchar xdata solcont[64]; uchar data solisp[20];
uchar solmsgbuf[0];
uint target,percent;

void tachometer (void) interupt 3 using 2 {
   union split{uint word;
           struct{uchar hi;uchar low;}bytes}};
   union split newcount;
   static uint oldcount,period;
   bb_enter_isr();
   bb_enable_event(updatereq);
   newcount.bytes.hi=TH1;
   newcount.bytes.lo=TH0;
   if (bb_event_occurred(updatereq)){
           /*tell control new value*/
      bb_ack_event(updatereq);
      period=newcount.word-oldcount;
      bb_put_message(ctl_tcb,&period,2);
   }
   oldcount=newcount.word;
   bb_exit_isr();
}

void control (void) {
   uchar len;
   uint period;
   forever{
      bb_wait(25); /*update every 1/4 Sec*/
```

The lines marked **1**, **2**, **3** appear to the left of `if (bb_event_occurred(updatereq)){`, `bb_put_message(ctl_tcb,&period,2);`, and `bb_wait(25); /*update every 1/4 Sec*/` respectively.

[1]The **updatereq** event comes from the control task to indicate that the newest data are desired. The event must be acknowledged or it will trigger the test repeatedly on each interrupt.

[2]The actual time difference is sent in the message to the control task.

[3]The $25 \times 10\text{mSec}$ time delay is to keep the update frequency fairly constant. Actually, if there is a long delay in the return of the latest tachometer reading, the time will be stretched.

Figure 11–6f   Speed Control—Byte-BOS

```
4       bb_post_event(&tac_tcb,updatereq);
                              /*request newest reading*/
5       if(bb_wait_for_message(&period,&len,10)==0){
            if (period < target) if (percent<100) ++percent;
            else if (percent>0) --percent;
        }
    }
}

void pulse (void) {
6   forever{
        P1=1;
        bb_wait(percent);
        P1=0;
        bb_wait(percent);
    }
}

void main(void){
    bb_init_bos();
    bb_create_task(&tac_tcb,&taccont,&tacisp,
                                &tacxsp,&tacmsgbuf,0,5);
    bb_create_task(&pls_tcb,&plscont,&plsisp,
                                &plsxsp,&plsmsgbuf,4,5);
    bb_create_task(&ctl_tcb,&ctlcont,&ctlisp,
                                &ctlxsp,&ctlmsgbuf,4,5);
    bb_run_task(&tac_tcb,tachometer)
    bb_run_task(&pls_tcb,pulse)
    bb_run_task(&ctl_tcb,control)
    bb_start_bos();
}
```

---

[4]Here the tachometer interrupt will be notified (when it gets around to checking) that a new value is desired.

[5]This wait for the reply has a timeout value to guarantee the wait will be no longer than 100mSec. Otherwise it resumes the 250mSec delay. This seems awkward and you may be able to come up with a better method. The heart of the problem is the fact that the interrupt is not a task and can't be awakened by another task.

[6]This is simplified from Chapter 9.

Figure 11–6f   Continued

```c
#include <reg51.h>
#include <rtxcapi.h>
#include <csema.h>
uint target,percent,period;
bit update;

frame tachometer (void) interupt 3 using 2 {
   union split{uint word;
               struct{uchar hi;uchar low;}bytes}};
   union split newcount;
   static uint oldcount;
   newcount.bytes.hi=TH1;
   newcount.bytes.lo=TH0;
   if (update){  /*tell control new value*/
      update=0;
      period=newcount.word-oldcount;
   }
   oldcount=newcount.word;
   return isrc(frame,updsema);
}

void control (void) {
   forever{
      KS_delay(0,25); /*update every 1/4 Sec*/
      update=1;   /*request newest reading*/
      KS_wait(updsema);
      if (period < target) if (percent<100) ++percent;
        else if (percent>0) -percent;
      }
   }
}

void pulse (void) {
   forever{
      P1=1;
      KS_delay(0,percent);
      P1=0;
      KS_delay(0,100-percent);
   }
}
```

The numbers **1**, **2**, **3**, **4** appear in the left margin beside the lines:
1 `if (update){`
2 `KS_delay(0,25);`
3 `update=1;`
4 `KS_wait(updsema);`

---

[1]Since the signalling from the interrupt is so primitive, it is necessary to use a shared variable (here a bit) to determine when a new reading is needed. In this way the math to determine the period need not be carried out until the result is needed—many unnecessary calculations would occur if the period is needed only once every 25 msec and the motor is spinning, say, 15000 rpm.

[2]The control algorithm updates every 1/4 sec so it can sleep (be blocked) the rest of the time. If precision is desired, it would be better to start a repeating interval timer and wait on the semaphore—that way the computation time would not be added to the 1/4 sec delay time.

[3]This signals the interrupt routine to produce a period calculation the next time around.

[4]This is an indefinite wait for the response from the interrupt that it is done. It would be possible to simply poll the update flag but that would require frequent waking (unblocking) of the control task to allow it to check the flag.

Figure 11–6g Speed Control—RTXC

Essentially you take the motor control outside the domain of the larger more cumbersome operating systems and do it yourself like the non-RTOS versions. On the other hand, you could use a D-A converter and choose a voltage to output directly. None of the "relatives" provide such on-chip, but the cost would not be high. To simplify the example, I'll assume that the *hardware* will smooth things out and the period can be 100×10mSec or 1Sec. This is a bit slow if the update interval for the control algorithm is 250mSec, so I'll assume a very large motor and update once a second for the big systems.

REVIEW AND BEYOND

1. What RTOS calls are involved in using an interrupt?
2. Why is the handling of an outside interrupt a special problem for some RTOS?
3. What is context switching and how is it most efficiently done in the 8051? How is it normally done in other computer systems with "flat" memory architecture?

Chapter 12

Distributed Systems

USING SEVERAL CONTROLLERS

When there are too many things to be done for one processor to keep up, it may be logical to go to *several* processors. Although major math algorithms can be sent out to a math chip and although faster processors are available—particularly the 8051's bigger relatives—it is possible that you will want to consider *several* processors linked by a serial communication scheme. For example, a robot could have one processor controlling the movement of the arm while another controller could handle the tactile (touch) sense in a gripper, and a third controller could provide interface with a vision system. In addition, there might be a master controller, which obtains general instructions from a factory computer system and converts them to specific commands for the robot.

On a smaller scale, a specific system is known to the author where one processor runs pumps and valves, a second one interacts with a user, and the third processor handles a fairly complicated measurement algorithm requiring significant amounts of processing.

In both cases, the job is partitioned so that individual processors handle *related* functions and only occasionally send small amounts of information between processors. Dividing work among multiple controllers is called multiprocessing or **distributed control**. This is distinguished from **parallel processing**, where a large amount of information is shared among closely coupled processors. In just the same way that you should partition tasks to minimize the amount of information flow, so you should partition distributed control systems so the information flow between controllers is as small as possible.

This chapter is only a survey and no specific examples are included. There is such a diversity of choices in the write-your-own class that it would be difficult to choose one. In the RTOS class, only BITBUS (DCX) and DCX-COMM have any direct distributed support and the changes are so small as to be negligible.

DIRECT PARALLEL

For a very close pair of processors, it is possible to directly connect two on-chip parallel ports. One processor could put out a byte of data and then send out a high bit on another port as a "data ready" signal. The second processor could pick up the data and set a port bit high as an "acknowledge" signal. For speed, these bits could go to each other's interrupt inputs. If data flows only one way, it might be possible to synchronize by sending a special code in between bytes of data (say a zero) or define a special character for the start character. Without handshaking, however, the process rapidly escalates in complexity and pulls in all the issues of serial protocols.

PARALLEL BUFFERED WITH HANDSHAKING

For distributed control there are at least two ways to proceed. First, it is possible to use ordinary serial (or parallel) communication between controllers. Parallel port chips such as Intel's 8255 have handshaking for bi-directional parallel communication, which would be appropriate for short-distance communication between two processors. One processor could leave off information in the port chip, which triggers an interrupt to the other processor. The second processor could also leave information off the other direction. The transfer would not need to be synchronized and could be with the O B F (output buffer full), A C K (acknowledge), and I N T R (interrupt request) signals. It could be done with only about 10 lines and two 8255 chips if mode 2 is used (leaving most of four ports free for other uses) or with a single 8255, 18 lines between processors and tying up two ports in mode 1. This method is seldom used for distant communication, but if you have several processors on one board, it would merit further study with closer attention to the 8255.

SERIAL INTERCONNECTION

With only two processors to interconnect, it is quite possible to simply tie together the serial ports—in-to-out and out-to-in. For short runs, the lines can be tied directly between processors with no buffers. If the distance is greater, then the addition of RS-232 or RS-422/485 buffers is possible.

If more than two controllers must talk, the issue of "who" talks "when" needs to be resolved for any communication scheme. It is a major concern of network protocols and hardware to avoid two nodes talking at the same time on the same channel.

SERIAL RING

A loop could be set up where each processor echoed incoming messages out to the next processor in the circle. This would lead to a token-ring sort of network where communication is initiated by the processor currently holding the token. Each processor around the circle would echo the incoming characters out to the next processor around the circle. The processor being addressed would receive the message and then send on around the circle either an acknowledge message or a reply message. When the reply got around to the token-holder, the task could send another message or forward the token on to the next processor around the circle. This is not a full study of networking, but it should be easy to see that extensive communication activity could tie up the processors.

NINTH-BIT TREE

If a tree structure is envisioned with a single "master" node and all the other processors only listening, then multiple listeners can be tied to one RS-232 line. Where more extensive serial communication is expected, the "ninth-bit" capability mentioned in Chapter 7 is useful. Then in a master/slave setting, nonaddressed controllers can all ignore (by disabling the serial interrupt) the serial transmission. They needn't waste processing time being interrupted to discard characters. Only the addressed slave will process the serial characters. When a new slave is to be addressed, setting the ninth bit can awaken all slaves to check the address.

RS-485

Beyond RS-232, RS-422 and RS-485 define a differential standard for communication where the single pair can support communication in either direction. Thus, one line can allow communication in either direction and any processor on the line can be the sender or the receiver without any hardware switching. That allows a tree with multiple "masters" as long as they don't try to talk at the same time. A token scheme is possible as long as the token doesn't get lost or some node will make a new one if the old one gets lost. Historically, this specification was first used with Intel's BITBUS, but Iota has recently introduced a system called DCE-COMM, which uses RS-485 without requiring the SDLC controller found on only the 8044. More details were not available at the time of writing, but it is expected that the system will behave much like BITBUS because DCE is much like DCX.

SDLC

The SDLC (Synchronous Data Link Control) protocol developed by IBM was adopted as the standard for BITBUS. It is like the scheme just described with a master and slaves in a tree structure, but in a block-structured protocol rather than a character-structured one with the ninth-bit scheme. A message block consists of a fixed arrangement of characters structure including a start code byte, an address, the block length, and a control code. After the variable-length block of data characters comes an ending structure of a CRC (Cyclic Redundancy Check) code and a fixed end code byte. There are considerably more details than it is appropriate for this book to cover, but the point is that the protocol guarantees a high degree of end-to-end message passing integrity and is supported by very standard interface chips. Intel set up the BITBUS system around the 8044 relative of the 8051, which has a SDLC controller on-chip.

I²C BUS

A versatile serial bus is the I²C bus protocol promoted by Philips/Signetics. The bus is well suited for very small systems (128 possible addresses) with relatively short, slow communication needs (100 kHz maximum clock rate). Its strongest features are its extreme simplicity and its multimaster capability.

The I²C protocol involves two lines (plus a ground reference) that provide the clocking (S C L) and the actual data transmission (S D A). The two lines are open-collector, bi-directional. The protocol works as a multimaster, multislave system where the master supplies the clocking signal. The master starts by sending an 8-bit address (7-bit address plus a bit to indicate write or read). The addressed slave responds by pulling the line low in acknowledgment (thus the master can detect the absence of a slave responding). Because the bus is open-collector with a single pull-up resistor, the availability of the bus is indicated by the lines being high. The start of a transaction is indicated by S D A going low when S C L is high. Data transitions occur only when S C L is low. The end of a transaction is indicated by S D A going high when S C L is high.

Although the bus is useful for interprocessor communication, there are a large number of intelligent peripheral devices available as well. A variety of voice and tone systems as well as several TV-related devices are available. Also, there are LCD drivers, D-A converters, RAM, EEPROM, and clock/calendar devices available. Thus, the bus is also an on-board substitute for parallel bus interconnection of peripherals where speed is not a serious concern.

BITBUS AS A NETWORK PROTOCOL

When Intel developed the DCX Operating System, they also developed a network protocol they called **BITBUS**. It defines a serial communication system or network oriented toward short control-sized messages. It is important to clarify that BITBUS is technically a *communication standard* rather than a set of hardware. The hardware part of the specification includes the use of simple twisted pair(s) as the transmission medium, RS-485 as the signal level and rate specification, and SDLC (Synchronous Data Link Control) as the standard for data blocks and error checking. The higher levels of the protocol include the message format that was mentioned with DCX messages. The production of BITBUS equipment is not limited to Intel; there are several companies that also make equipment for the BITBUS protocol and are discussed more in a later section. Several U.S. industrial control system companies also make similar proprietary systems, but there apparently is more interest in BITBUS in Europe than in the United States.

The BITBUS standard specifies the maximum number of nodes and total distance for the network. Three baud rates, 2.4MBaud, 375kBaud, and 62kBaud, are possible, although the fastest rate requires a separate clock line and is valid only for shorter distances. With repeaters, the slowest rate is good for 256 nodes and about 10 miles! Obviously, it can be quite a robust system for industrial control settings.

BITBUS AS A SYSTEM

BITBUS does have several drawbacks. An 8044 is almost mandatory, because no other 8051 family member has the necessary SDLC controller *and* shared memory that are the heart of the message scheme. All the messages between nodes must use the allocated buffers in on-chip RAM. At the same time, having the 8044 dramatically reduces the communication load on the processor by allocating the communication to the on-chip SDLC controller. The main processor's communication task simply leaves off a message in shared on-chip RAM and signals the SDLC controller side of the chip to handle it.

BITBUS is a strict master-slave system. The slave nodes should "speak only when spoken to." There is no collision detection. If a node initiates a message when the bus is in use, both messages will be corrupted. The system may crash in the sense that nodes may have to be reset, but there will be no hardware failure as a result. There is no way for a slave node to signal the master if it has some urgent message, so the software ought to be designed so the master node at least polls the slaves periodically to see if they have any messages to send. Likewise, slaves cannot communicate directly with other slaves—messages must be relayed by the master node.

Although the communication rate is much faster than RS-232, it is much slower than Ethernet or other data-oriented protocols. The ratio of actual message to total bits sent is not high—about 50 percent by the time the message header and the SDLC overhead are included. For long messages such as downloading programs, the message is broken into many small pieces. However, the protocol is well suited for short messages such as control systems normally use.

Even with the drawbacks, the simplicity of adding BITBUS to the DCX is impressive. Moving tasks off to other boards when a system grows too large only requires changing the node and task identifications in the message header. Otherwise, assuming the original tasks avoided shared variables and using on-chip RAM for messages, there is no change at all. This is one of the strengths of DCX and BITBUS.

IF YOU HATE TO PROGRAM!

Apparently, many BITBUS users do not actually distribute the control software. Instead, by using a preconfigured task called the RAC (Remote Access and Control) task, they use the nodes as remote "dumb" I/O with a low-cost serial link from the host controller. Although this is a far cry from distributed *control*, it apparently is quite easy to get going and requires little programming skill. The *FILBUS* from Gespac has a strong emphasis on the ease of use by less software-oriented people. It has extensive preconfigured task software built into the nodes to the extent that they can guess what would be useful. There are entire timed sequencer tasks for digital I/O boards as well as moderately fast analog-output sequencing tasks controlling levels and slopes of the signal in a repeated manner without further remote intervention. Data collection is also available in preconfigured tasks. There are even boards and tasks for keypad, LCD display, and CRT terminal interface. Both Intel and Gespac have interface cards to PCs so the entire distributed control system can be addressed from a high-level host programmed in, for example, C. They also have interfaces to high-level *multitasking* hosts OS-9 and RMX-86 respectively. This again goes beyond the scope of this book.

PARTITIONING FOR DISTRIBUTED CONTROL

In partitioning tasks among distributed nodes, remember that SDLC protocol requires a master/slave relationship, so messages should not be *originated* by tasks on a slave board to tasks on any other board. A master node is usually set up to shuttle messages between nodes by sending a message to each node that invites a reply (or a "no messages now" indication). Because such a process is somewhat slow and cumbersome, it is advisable to keep closely coupled tasks together on one node and put unrelated tasks on other nodes.

What determines the master node? Actually, there is no hardware involved, and it is a matter of agreement and convention. All the slaves must agree to act like slaves. Messages could be originated anywhere, and the operation would go smoothly until two nodes tried to talk at once. It would be possible to arrange a transfer of masters as long as it was agreed upon in the software. Also, a token passing system is possible in which the master could send the token to a slave to permit it to become the new master. The main problem is that such an approach would not be compatible with the convenient debugging tools that Intel, for example, provides for its BITBUS products.

REVIEW AND BEYOND

1. What differences are there between multitasking and distributed processing? Why can DCX support both quite easily?
2. What can be done if there is not enough total CPU time to get all the tasks done?
3. How is *distributed* processing different from *parallel* processing?
4. Discuss in detail some aspect of an automobile that could involve distributed control.
5. Compare the various serial communication schemes discussed.
6. For a very close 2-controller system, compare the benefits and drawbacks of parallel and serial communication.

Chapter 13

Multitasking in Review

BENEFITS OF RTOS

Having described multitasking *real-time operating systems* (RTOS) along with *distributed control operating systems*, it may help to reconsider the traditional monolithic program and the "write-your-own" multitasking approaches such as the scheduler of Chapter 9 and the ninth-bit communication of Chapter 12. Given enough programming effort, they have many of the same features as the operating system. They may even run more efficiently. There are no license fees to pay or any initial purchase costs. They can be quite small and can (perhaps) be totally understood by you, the programmer. All the code is visible. What are the reasons to go to RTOS?

Real-time operating systems are an aid in *formalizing* the relationship between activities so that the software development can be simplified. RTOS are particularly valuable for projects in which there are *several* programmers working over a long period of time with the likelihood of frequent revisions of pieces of the project. That probably describes most microcomputer projects!

The breaking up of a project into functional pieces called tasks allows modular development of the software much like that described in Chapter 1. In addition to the strictly hardware examples used here, data processing and reduction can fit a modular multitasking model well. For example, one task might be to take a block of raw input data (from an A-D converter) and pick out the high and low values. That might in turn give rise to an alarm message sent to another task that would take corrective action. Additional processing might entail digital signal processing to determine frequency components or to adjust process variables to maintain a desired set point. All of this *could* be done in one block, but it is the *formalizing* of things that helps when several people are involved in the project.

When jobs are broken into tasks, it becomes easier to grasp the overall picture without becoming lost in the details. In addition to formalizing the process, the operating system can make the interrelation between tasks more comprehensible to someone else. With a formal relation between tasks, there will be less chance of overlooking or forgetting details. When the inevitable revisions come, there will be less start-over-from-scratch work.

A second benefit is the simplicity of handling *real-time* inputs. Either write-your-own or operating-system multitasking can make it easy to respond to real-time demands compared to the single monolithic program approach. If the keys need to be scanned every 40mSec, a task can be designed to sleep just that long, no matter what else is going on. If the display *ought* to be updated every 100mSec, that can be arranged with a lower priority because it is not mandatory. The operating system does not perform magic in the sense that more than 100 percent of the processor's time becomes available, but it makes it possible to clearly define the different activities and the degree of urgency (priority) of the activities. One still faces the *"What if this needs to happen when that is going on?"* questions, which are part of any real-time system. But, compared with single monolithic programs, the solutions are more obvious to someone else, and are not buried in the interrelation of large code modules.

Third, *intertask communication* (signals and messages) provide a solid method of controlling execution order and timing. In other words, it is possible to move from one activity to another in a way determined by the events that have happened. The sequence is not locked in, as it might be with a scheduler you would write yourself. At one time a message from one task can be sent to start a second task running, but at some other time the message might be sent to a third task instead. If you are using BITBUS, the message scheme is not much more complex, and you can readily move to distributed controllers.

Fourth, tasks can be *small*, making them easier to manage. They can be *independent* so some of the interaction details of single-program solutions are avoided.

Finally, BITBUS allows the controllers to be located near the things to be controlled with a corresponding reduction in message traffic. When a node fails, the entire system has not failed. The individual nodes can be programmed to take corrective action directly much like you respond to a hot stove before the message has traveled all the way to the brain.

COSTS OF RTOS

What does multitasking *cost*? Some of the costs have been alluded to already. The hardware must have a specific timer and interrupt structure. That is why, for example, DCX will not run on a PC. The DCX 51 operating system was designed specifically for the 8051 family of chips, which have the necessary hardware in the chip. Versions of some RTOS exist for other processors, but there must be certain hardware in those systems and the RTOS must be configured to match the specific hardware there.

In addition to the hardware requirement, multitasking costs *time*. The *overhead* of all the operating system activity uses up computer time. Periodic interrupts to update tables costs time. Sorting through to determine which task should run takes time. If the task relationships are very simple and inflexible, it might be better to write a monolithic program. Of course, if the processor is so heavily loaded that the overhead is a problem, it may mean that a different approach is in order— a faster processor, multiple processors to handle different pieces, or specialized devices such as math co-processors.

A third very real cost is the time for the application programmers to get up the *learning curve* far enough to become productive. Again, for a single very small project, multitasking probably doesn't make sense. But once a group is familiar with the concepts, the formalization and modular nature of the process make for much more *maintainable* code. It is the same problem faced in making a transition from assembly language to a higher level language. Is it worth the time and pain to learn? The answer depends on one's particular situation. Having a taste for some of the RTOS from the previous chapters may help you come to a decision. You have seen that DCX, RTXtiny, and DCE have very few commands to learn. CMX Byte-BOS, RTXC, and USX are much more complicated in one sense, but in another sense they leave less to be done by you because the careful use of the different system calls can avoid the need for several other system calls. There will be no substitute for study!

Appendix A

8051 Assembly Instructions

The section that follows summarizes the assembly language mnemonics used by Intel (Copyright 1980, but other 8051 assemblers seem to use the same mnemonics anyway). More detailed information is available in the "MCS-51 Programmer's Guide and Instruction Set" of Intel's *8-bit Embedded Controllers* data book. This appendix is designed to help you choose instructions or to understand their function in a listing. The data book has a nice numeric listing of all the machine codes, which is useful for manual disassembly. It also has a more detailed description of each machine instruction.

A WORD ABOUT ACCESS AND ADDRESSES

The instructions reflect the hardware design. There are 8 registers that start at address 0 (for `registerbank` 0). There can be 4 sets of registers all with the same `R0` through `R7` designation. They will be found at absolute addresses 0–7, 8–F,10–17, and 18–1F. Which bank is being referred to depends on two bits in the program status byte (`PSW`). Any unused registerbanks as well as the rest of the 128 (or more) bytes of internal memory are available for program use as long as it doesn't interfere with the stack or with the bit-addressable section. All of the lower 128 bytes of RAM and the special function registers (`SFR`) are directly addressable.

`SFR`s of particular interest are the accumulator (`ACC`, at address E0), the E register (at F0), and the data pointer (`DPH` at 83 and `DPL` at 82). They all figure prominently in the instructions.

In addition to directly accessing the 128 bytes of RAM and the S F R s, R0 and R1 can serve as pointers to any RAM above 128 (as in the 8044, 8052 and many other family members).

There are a set of locations that are addressable as single bits from 20h through 2fh addressable from 0 to 128. In addition, some of the S F Rs are bit addressable in the range from 128 through 255.

Finally, the D P T R is the only pointer to off-chip RAM and code space. There are only three instructions for such access and they are discussed in the move instructions.

DATA MOVING INSTRUCTIONS

mov

The move instruction has a host of different forms depending on where the data comes from and where the data goes. A m o v does not destroy the data in the source—rather it copies it to the destination. A few combinations are not possible. It is not possible to move register to register, but once you realize that for a given register bank each register also has a direct address, you can get at them that way. There is also no indirect/indirect move. The kinds of moves are listed below with examples.

Accumulator/register	mov a,r7	mov r1,a
Accumulator/direct	mov a,22h	mov 03,a (3 -> r3 in bank 0)
Accumulator/indirect	mov a,@r1	mov @r0,a (r0 & r1 only)
Accumulator/data	mov a,#22	
Register/direct	mov r3,7fh	mov 6eh,r2
Register/data	mov r1,#5fh	
Direct/direct	mov 1fh,7eh	
Direct/indirect	mov r3,@r1	mov @r0,r6
Direct/data	mov r7,#01	
Indirect/data	mov @r0,#7fh	
Data pointer/data	mov DPTR,#0ffc0h	(2-byte load)
Bit/carry	mov c,acc.2	mov 20.3,c

Notice the significance of the **#** and **@** symbols. It is very common to forget the **#** when intending to move data—the result is a fetch from some direct address, which may be holding anything! It is quite common to use equates in place of the direct addresses so the moves might look like, m o v relaystatus,#0ffh. Also, the special function registers are usually referred to by name (already known to the assembler) so you would write, m o v CON,#013h.

movc

Access to code space (usually EPROM) is by definition read-only. Some hardware will OR the PSEN and RD lines to produce overlapping addresses in RAM so downloading of code from a host development system is possible, but the concept is still one of read-only for code. The DPTR move is quite useful for the fetching of data from ROM-based lookup tables.

movc a,@a+dptr (note this destroys the value of a)

movc a,@a+pc

I have never seen the latter instruction actually used, but it presumably could be used to fetch values from a table stored just following the instructions as shown in the detailed instruction discussion in the Intel data book.

movx

This is the mainstay of most large programs because no operations work directly on external RAM and most applications have become too large to work with less than 128 bytes of storage. The page movx using r0 or r1 is quite uncommon because it only puts out the 8-bit address on P0 and doesn't affect P2. If you use lots of ports and no large external RAM, it could handle the I/O addressing. Alternately, you could put out a page select on P2 just before using this instruction. These approaches are for systems where lack of a few port pins would destroy the cost savings of a minimal system and are generally not used.

movx a,@dptr   mov @dptr,a

movx a,@r0   mov @r1,a

These are the standard instructions for addressing external I/O ports as well as other variables. With high-level languages, this is transparent to the programmer.

xch

Unlike the mov that copies from one place to another, the xch swaps the two bytes. It is particularly because many operations involve the accumulator

Accumulator/register   xch a,r4

Accumulator/direct   xch a,1eh

Accumulator/indirect   xch a,@r1

push

This puts a byte onto the stack. Any of the direct addresses can be pushed, including the S F R s. Thus, it is possible to push the accumulator, B, P S W, and the various hardware control registers. It is not possible or necessary to push r 0 through r 7 by name because it is preferable to switch registerbanks by changing two bits of the P S W.

push B

The details of context switching with interrupts are discussed in Chapter 7, but most of it is transparent to the high-level language user.

pop

This is the reverse of the p u s h. Remember that the sequence should be the reverse of the pushes unless you wish to get things crossed up.

pop PSW

BRANCHING INSTRUCTIONS

There are some very powerful branching instructions for the 8051. There are actually three different addressing methods for jumps and calls, but most assemblers will handle it for the programmer. In fact, many assemblers will accept j m p and c a l l as the instruction and determine the most efficient instruction to use. The short jump covers an address range of 128 bytes back to 127 bytes on down from the instruction. The a j m p and a c a l l supply the lower 11 bits of the 16 bit address and keep the upper 5 bits of the next program instruction. This forces the destination to be in the same 2K block as the call. It makes a nightmare for the linker designer or the assembly programmer when relocatable code is involved, but it does save a byte on the addresses. Finally, there are the l c a l l and l j m p that include the full, absolute 16-bit address of the destination.

Unconditional jmp

These are jumps that occur without testing.

a j m p subroutine (must be in same 2K block)
l j m p po in t a (anywhere in 64K code space)
s j m p wa i t i ng (relative jump +127 to −128)

Conditional jmp

These test and either make the short jump or else flow on to the next instruction.

jz pointx (if accumulator is currently all zero)
jnz pointy (if any bits of accumulator are not zero)
jc pointz
jnc pointz
jb P3.5,pointa
jnb P3.1,pointb
jbc 22.3,pointc this also clears the tested bit

cjne

Compare and jump if not equal. Note that not all combinations exist. Also note that the carry flag is set as a result of this instruction, so it is often more useful as the first part of a greater-than/less-than test.

cjne a,3eh,pointz
cjne a,#10,pointw
cjne r5,#34,loop
cjne @r1,#5,goingon

You can only compare a register with the accumulator by using the register's direct address, which depends on the registerbank.

As an example of the inequality test, here is a code piece where, say, the value of r6 should determine whether program flow should branch to GR, LE, or EQ depending on whether r6 is greater than, less than, or equal to 20:

cjne r6,#20,NE
EQ:...
jmp ...
NE:jc LE
GR:...
jmp ...
LE:...
jmp ...

djnz

This is a very handy instruction for iterative loops where the number of times to go around is set outside the loop and then counted down to zero.

```
djnz r3,pointq
djnz 3fh,pointj
```

call

As mentioned, there are 2 addressing ranges. The call pushes two bytes onto the stack—the address of the next instruction following the call so the return can set the program counter (PC) back to the place to resume after the subroutine.

```
acall stepperroutine
lcall display
```

ret

The returns put the top two values on the stack into the program counter and allow the flow to resume following the completion of a subroutine. The reti additionally restores the interrupt logic to allow further interrupts of the same priority level.

```
ret
```

```
reti
```

It is possible to play games with the call and return instructions to not return to the place where the call originated, but it is dangerous! It is used in the DCX operating system to handle task switching, because, if the program alters the top two bytes on the stack, the flow will return to a different place.

ARITHMETIC OPERATIONS

All of the arithmetic is 8-bit and is discussed in Chapter 2. Other than increment and decrement, all arithmetic leaves the result in the accumulator and destroys the previous accumulator contents involved.

add

This doesn't include the carry bit coming in with the least significant bit (lsb), but it does produce a carry result. It is a choice as first step of a multibyte operation, but many prefer to zero the carry bit and use the addc throughout the loop.

```
add a,r5
add a,22
```

```
add a,@r0
add a,#22
```

addc

This includes the carry bit in the addition.

```
addc a,r5
addc a,22
addc a,@r0
addc a,#22
```

subb

There is no subtract without the borrow. If you know the normal 2's complement subtraction by addition, see the borrow as the *absence* of a carry out. Otherwise, consider the borrow, which is the carry flag, as the indication that something too big was subtracted. For multibyte subtraction, that is OK as long as there is no borrow at the end of the process. Remember that this is unsigned math and any adjustments for negative numbers or results will have to be made in software that you write.

```
subb a,r5
subb a,22
subb a,@r0
subb a,#22
```

mul

There is only one hardware multiply, which produces a 16-bit result in the accumulator (low byte) and the B register(high byte). If the product exceeds 8 bits, the overflow flag is set. The carry flag is always cleared.

```
mul ab
```

div

The accumulator is divided by B with the result in ACC and the remainder (not the fraction) in B. The carry and overflow flags are always cleared. Neither operation neatly supports multibyte extensions.

```
div ab
```

inc and dec

These are symmetrical except for the data pointer. Decrementing 0 or incre-menting f f will fold over to f f and 0 respectively.

```
inc a           dec a
inc r2          dec r5
inc 45h         dec 3eh
inc @r0         dec @r1
inc dptr
```

LOGICAL OPERATORS

The logical operators allow bitwise manipulation of variables and are discussed in Chapter 2 in some detail.

anl

Logical ANDing produces a high where both bits are high. Notice that this instruction does not leave the result in the accumulator!

```
anl a,r6
anl a,25h
anl a,@r1
anl a.#03h
anl 25h,a
anl c,acc.5 (bit operation)
anl c,/acc.5 (use complement of bit)
```

orl

Logical ORing puts a one in a place if either bit is high.

```
orl a,r6
orl a,25h
orl a,@r1
orl a.#03h
orl 25h,a
orl c,acc.5 (bit operation)
orl c,/acc.5 (use complement of bit)
```

xrl

Logical E X C L U S I V E O Ring puts a 1 if one and only one bit is high—if both bits are high it puts a zero.

```
xrl a,r6
xrl a,25h
xrl a,@r1
xrl a,#03h
xrl 25h,a
```

cpl

The complement puts a zero for a 1 and a 1 for a zero.

```
cpl a (all 8 bits of the accumulator)
cpl c (the carry bit)
cpl P3.5
```

clr

This clears the byte or bit involved. The bit operations work on the 128 bits in the 20h-2fh area as well as the bit addressable S F Rs including the accumulator and the on-chip ports.

```
clr a
clr c (the carry bit)
clr acc.7
clr P1.5
```

set

The set makes the indicated bit a 1.

```
setb c
setb 20.3
setb acc.7
```

rotates

The byte is shifted by one place left (toward msb) or right (toward lsb). If the carry is included, then the end bit goes into the carry and the carry goes into the other end.

r l a

r r a

r r c a

r l c a

DECIMAL INSTRUCTIONS

These few instructions are almost always used with binary coded decimal (BCD) data and are put here to emphasize that point.

xchd

This instruction exchanges the low-order nibble of the accumulator with the indirect addressed value. The upper nibble remains with the original location.

x c h d @ r 0 , 0 2 a h

swap

This instruction reverses the places of the upper and lower nibble of the accumulator.

s w a p a

da

This decimal adjusts the accumulator. If earlier addition of packed BCD digits was done where the A C flag was set or the current values of either nibble exceed 9, then this instruction adds 00h, 06h, 60h, or 66h as needed to bring the digits back into decimal form.

da a

Note that this will not do a magic hex-to-decimal conversion and is of no use for decimal subtraction. In general it is preferable to do math in binary (hex) form and only convert to decimal when user input or output are involved.

Appendix B

Language Details

LANGUAGE SWITCHING HINTS

8085 Assembly to 8051 Assembly

There are several differences that should make the transition easier. First, there is no I/O space in the way that the 8085 defined it. There are the internal ports, which are defined as registers, but any added hardware is addressed as off-chip (external memory). There is only one way to get to it and that is by the movx command. The internal registers are R0 through R7 as well as ACC and B. Other than loading DPTR, there are no 2-byte loading instructions—it is much more an 8-bit machine.

The @ symbol means indirect. Only DPTR, R0, and R1 can be used for such moves. Be careful to see that mov a, r0 is not the same as mov a, @r0. The former brings the contents of r0 into the accumulator, whereas the latter brings the contents of the place where r0 points into the accumulator. Another easy error is the omitting of the # sign. mov a, #25 is quite different from mov a, 25; the former puts the binary equivalent of decimal 25 into acc, whereas the latter puts the contents of address 25 (19h) into acc. Notice that all these moves work only with internal RAM—there is only one way to get external RAM values—the movx.

With a good assembler, the different kinds of calls and jumps can be ignored. If you use call and jmp, the assembler will pick the smallest sjmp-ajmp-ljmp-acall-lcall that will reach.

Notice that the zero flag, reflects the current state of the accumulator at all times—you do not have to do a comparison instruction to get the flag set. But there is only one comparison operation available. The cjne is quite powerful because it not only branches for inequality but it also sets the carry flag for the first byte being less. The d j n z instruction makes looping quite efficient.

The only way to access code (EPROM) space is with the m o v c instruction, which is obviously set up for accessing look-up tables. It is impossible to write to code space unless a hardware O R of the P S E N and R D lines is included in the hardware.

The m u l and d i v instructions are new. They are good for only 8-byte operations and it is not clear if they do much for multibyte math that could not be better done in more traditional ways.

PL/M 80 to PL/M 51

Very little changes between these two languages. The most obvious change is the loss of the I N P U T and O U T P U T instructions.

That is a natural result of the hardware change. Because all I/O is memory mapped, it is easy to D E C L A R E any port with the A T ( ) to give it whatever name is desired. Along with D E C L A R Es, the choice of memory type is needed through use of the words A U X I L I A R Y , M A I N , B I T , R E G I S T E R, or C O N S T A N T. The word D A T A is now replaced by C O N S T A N T, because all the data tables go into ROM space.

In the actual program space, the only differences I have noticed relate to the forbidding of embedded assignments (the : =) and an occasional complaint when too much is put in one line. The possibility of B I T variables will be new and the compiler will not let you use them like 1's and 0's in math operations. You will notice that the 2-byte values are stored in reverse order from PL/M 80 (or 86).

PL/M 86 to PL/M 51

Again, very little changes between languages. The same memory space comments of the previous section apply. The biggest shortfall in PL/M 51 is the lack of support for any variable types but B Y T E and W O R D. This is the reason that some applications went to C for the 8051. A few of the string operations are not available as well, but the general experience is that nothing else has changed.

PL/M to C

This transition is not as difficult as it might seem. From a study of Chapters 2–5, it is obvious that the structures are mostly the same and that anything you can do in PL/M can also be done in C. Because the order of a few structures is reversed, it is not possible to do a totally automatic conversion with simple replacement. It *is* possible, however, to go most of the way with substitutions and then do some manual fix-up. Then it is possible to condense the C a bit more with assignment operators and embedded assignments in f o r and i f loops, but the result (without the condensation) is quite valid C and is probably more easily understood. It becomes a matter of style and choice, although the condensed version may compile to slightly tighter code in some cases.

Table B–1 lists some of the substitutions and may help PL/M programmers to change over. Remember that parentheses for arrays must go to square brackets and `if` and `for` blocks require parentheses. When you make multiple assignments to a common value (setting several variables to zero, for example), in PL/M you use commas while in C you use successive equals. The comma in C has only a few uses such as in a `for` loop where you wish to do two things where you normally have only one. Where you declare several variables of the same type in PL/M, you surround them with parentheses, while in C you separate them with commas.

TABLE B–1 PL/M to C Substitutions

`DO;`	`{`
`end;`	`}`
`FOR`	`for(`
`IF`	`if(`
`THEN`	
`ELSE`	`else`
`DO WHILE`	`while(`
`DO CASE`	`switch(`
`=`	`=` or `==`
`<>`	`!=`
`AND`	`&` or `&&`
`OR`	`|` or `||`
`NOT`	`~`
`XOR`	
`MOD`	`%`
`BYTE`	`unsigned char`
`WORD`	`unsigned`
`MAIN`	`data` (Franklin C)
`AUXILIARY`	`xdata #`
`CONTSTANT`	`code #`
`LITERALLY`	`#define`

Standard C to C51

Much of this is discussed in the main part of the book. The changes you will want to include involve additions to C—mostly they specify the sort of memory space you want used for variables. The memory model directives allow you to pick a default treatment.

You are probably used to having a console/printer and may miss the `printf()` and `scanf()` functions. If you consider the normal 8051 applications, there *is* no standard and may not even be any device for such I/O! Most of the C51 compilers allow you to supply one or two primitive drivers to go either to an LCD module or a terminal (via the serial port) and come from either a keypad or a terminal.

But for many embedded applications, there will ultimately be no terminal involved and it is crazy to try to make an 8051 into a personal computer!

Basically, the biggest change you face in switching languages is the change in *thinking*. It takes experience to avoid producing huge, inefficient programs due to requests for float and trig functions. It is necessary to think in terms of simpler math or precomputed look-up tables rather than coordinate transformations, and so on. I believe the best C51 programmers probably move up from assembly rather than down from standard C.

C COMPILER INFORMATION

The information that follows is the work of Charles Larson, which he has kindly adapted from his article in *Embedded Systems Programming,* "8051 C Cross Compilers" (May 1991, pp. 49–62). The only known omission is the compiler from 2500AD Software which declined to have its product evaluated for the magazine article. It seems to be relatively weak for support and help to new programmers and requires the user to write assembly start-up routines. Note that the article is from 1991 and the field seems to be changing rapidly—do not assume that any of the specific complaints will still be applicable without checking with the individual vendors. Check Appendix D for addresses and phone numbers.

The 8051 is a popular microcontroller. It contains a CPU, serial port, parallel I/O, two timers, two external interrupts, ROM, and 128 bytes of RAM, all integrated into a single chip that can be purchased for as little as $1.50. The 8051 is available from a number of manufacturers: Intel, OKI, AMD, Philips/Signetics. Siemens, and others. Many different kinds of derivative parts are available that use the basic 8051 architecture augmented with special functions, such as analog-to-digital converters, additional timers, input-compare and timer-capture registers, and so on.

The most common complaint about the 8051 by many users is that its assembly language is somewhat convoluted and difficult to work with. For example, only two instructions are provided for accessing the 8051's program memory (as might be required to store an ASCII string for a display prompt). One instruction is program-counter relative, the other requires the use of a data pointer and offset. Both instructions write the data over the offset value. An operation as simple as integer addition requires a minimum of six instructions—more if the operands happen to be in external memory. A good programmer can always work around these limitations, but will obviously work more efficiently if these problems can be handled transparently.

The solution to this problem is to use a high-level language, such as Forth, Basic, C, or PL/M. Of these, I think C gives the best combination of readability, portability, and access to the processor's features. I can usually finish a project written in C in less than half the time it takes to write it in assembly language, although quite a few assembly language programmers I have spoken with don't seem to be convinced. The high degree of portability means more than just the ability to move the code from one processor to another. This degree of portability also means the programmer won't have to spend a great deal of time figuring out the language or even the peculiarities of a particular dialect, which is especially important since a significant portion of any software project is code maintenance, and quite a variety of programmers may have to work on the code.

Although typically I use C when writing PC-based programs, I used to think the limitations of a microcontroller made C impractical. About a year ago, however, I wrote an 8051 program for a client who wanted it written in C. My experience with the project convinced me that not only was C practical, but that is was the language of choice. The program was a laser-tube controller that contained a keyboard and display interface, a state machine for control, two interrupt routines for a timer and phase-controlled power, and diagnostics. The entire program required about 1,500 lines of C source code, and the compiled code fit into the internal EPROM and RAM of an 8752. Not a single line of assembly code was used, and the project was completed in three weeks.

Once I found that C was practical on the 8051, I decided to find the best compiler for it. The compiler was of particular importance to me because I represent a line of in-circuit emulators for the 8051 and wanted to be able to make accurate recommendations to my potential customers. I found eight compilers to review. Six (produced by American Automation, Archimedes, Avocet, BSO/Tasking, Franklin, and Intermetrics) were complete near-ANSI compilers. I found two other, more limited packages: one from Micro Computer Controls, and a shareware package written by Dave Dunfield.

8051 ARCHITECTURE

For those of you who haven't used the 8051, some explanation of its architecture is necessary to appreciate the compiler features required for effective code generation. The 8051 has three address spaces, program memory, internal data memory, and external data memory. The internal and external data memory can be accessed in two ways. You can access 128 bytes of internal memory (DATA) directly and indirectly. Some chips also have another 128 bytes of internal data, which is only accessible indirectly through the use of an 8-bit pointer register. These 256 bytes are collectively referred to as IDATA. The external memory, XDATA, may be as large as 64 kbytes and can only be accessed indirectly using a register called DPTR. The 8051 also has an instruction that treats the external memory as a 256-byte space, referred to as AUXPAGE. This space is useful in several processors that have 256 bytes of external memory on chip. The only 16-bit operation available is used to increment the DPTR register. All arithmetic must be done through the A register, and the only built-in stack is limited to the 256-byte IDATA space.

To cope with this memory structure, the compilers generally use one of the memory models shown in Table B–2. While the implementation varies from vendor to vendor, the models can generally be differentiated by the way local and global data is stored. For example, some compilers require function arguments to be passed on the internal stack regardless of the memory model. Others simulate the stack in external memory in the models supporting XDATA.

Since the stack-addressing capability of the 8051 isn't very flexible, some compilers use a technique called a compiled stack. Using a compiled stack, values such as function arguments and automatic variables, which would ordinarily be pushed on the stack, are allocated as static variables. The order of function calls is analyzed to determine which functions can be active simultaneously, and variables are allocated overlapping areas in

TABLE B–2 Memory Models for the 8051.

Model	Code Space	Locals	Globals
Tiny	64 kbytes	DATA	DATA
Small	64 kbytes	IDATA	IDATA
Compact	64 kbytes	DATA	XDATA
Medium	64 kbytes	IDATA	XDATA
Large	64 kbytes	XDATA	XDATA
Banked	>64 kbytes	XDATA	XDATA
Auxpage	64 kbytes	AUXPAGE	AUXPAGE

memory. This procedure takes up the same amount of room as a stack would at its maximum depth, but doesn't require the additional overhead of stack operations and the indirect access of variables—a particularly complex operation on the 8051.

BENCHMARKS

In choosing a testing approach, I tried to focus on how the 8051 is best used. Since it's a microcontroller, it is most likely to be used in cost-sensitive applications. It is a very efficient processor for writing control operations, as long as all the data memory is internal. When external data is required, the code becomes much more complex, and the system cost increases. These factors helped me to conclude that small-model performance is the most important test of the compiler.

I also consider math support (float and long) to be very important. In this situation, size is more important than speed. (If the project requires a lot of fast math, the 8051 would be an inappropriate choice.) On the other hand, faster development can be more important than small differences in code size—one or two simple calculations are much easier to code using a float or long. The time savings of having them built into the compiler is significant. Having a floating-point library available is almost worth the price of the compiler. As an example, I just completed a project containing five lines of floating-point instructions. The project could have been done with the same number of integer-arithmetic instructions, but the order of operations would have been critical and each calculation would have required a tediously calculated scaling factor as well as the extra effort necessary to comment the code so it could be understood later. Having floats available cuts a four-hour project down to 30 minutes.

I have summarized some general features in Table B–3. To get a quantitative figure on each compiler's performance, I ran seven benchmark programs on each compiler. Five were targeted at specific aspects of the compiler: integer, long and floating-point arithmetic, arrays, and pointers. The other two programs were more general—the sieve benchmark and a general arithmetic test. If, for some reason, the program would not run under small model, I used large model.

The code was compiled on a 12MHz 286-compatible with a 40-Mbyte hard disk. The execution was timed in cycles, using a Nohau emulator with an 80535 pod. At normal 12Mhz

TABLE B–3 General Features.

	Version	Compilation Time	Memory Models	Compiled Stack?	Float Support?
American Automation	16.02.07	6:03	SML	No	1
Archimedes	4.05A	2:03	TSCMLB	Yes	Yes
Avocet	1.3	1:47	SML	No	Yes
BSO/Tasking	1.1D	2:25	SAL	Yes	2
Franklin	3.01	1:28	SAL	Yes	Yes
Intermetrics	3.32	2:52	SL[3]	No	Yes
MCC	1.7	—[4]	SML	No	No[5]
Dunfield	2.11	—[4]	SL[6]	No	No[5]

Notes:
1. In large model only.
2. Available April '91.
3. Several static allocation schemes supported.
4. Could not compile all test programs.
5. Significant other limitations, see individual review.
6. ROM and RAM must be mapped to the same address space.

8051 clock speed, one cycle generally corresponds to 1 microsecond. The results of these tests are summarized in Tables B–4 and B–5. To give a better idea of the quality of code generation for the various compilers, I've shown the size in two tables. The size of the start up and library code varied somewhat, as can be seen in Table B–4, which lists total program size. The code size in Table B–5 shows the module size for the test program. The execution times shown in Table B–6 show the run time of the test modules without the initialization code.

Benchmarks are frequently maligned for not being accurate measurements of the compiler's performance. On some compilers and processors, the same algorithm can be recoded to show better results for different compilers.

The information that can be derived from the performance figures of a compiler is limited. To look at a benchmark as the absolute gauge of a compiler's performance is a mistake. Frequently, one compiler will perform better than another on one benchmark and worse on the next. The only benchmark that really counts is your application. The best these numbers can do is to give you an idea of the relative performance of each compiler. Keep in mind that a 10% change in code size is only going to make a difference if you are already using 90% of your available code space. Similarly, a 10% change in execution time only makes a difference if you are using 90% of your processor's instruction bandwidth.

As the numbers get closer together, the importance of performance diminishes and is overshadowed by other concerns. How easy is the compiler to work with, and how often does it generate unexplained problems? How easy is it to find answers to your problems in the documentation, and how responsive is the support staff at the company?

There are several additional factors to consider. These factors include quality of the

TABLE B–4 Total Program Size for the Test Programs.

	Tint	Tlong	Float	Array	Pointer	Arith	Sieve
American Automation	1,697	2,345	2,334	722	571	2,899	427
Archimedes	937	1,520	1,680	368	500	1,548	335
Avocet	1,067	1,849	1,657	393	368	1,250	296
BSO/Tasking	1,223	1,690	—	512	413	1,534	492
Franklin	776	1,343	1,390	181	322	816	166
Intermetrics	1,065	1,407	1,426	358	277	875	255
MCC	2,802	—	—	990	—	—	602
Dunfield	1,605	—	—	854	736	—	817

TABLE B–5 Module Size for the Test Programs.

	Tint	Tlong	Float	Array	Pointer	Arith	Sieve
American Automation	1,234	913	410	380	229	646	171
Archimedes	654	1,027	349	183	153	478	150
Avocet	720	1,238	567	263	270	612	166
BSO/Tasking	650	784	—	190	91	326	170
Franklin	614	979	363	166	156	489	151
Intermetrics	715	718	349	212	131	415	154
MCC	2,230	—	—	517	—	—	315
Dunfield	1,189	—	—	437	319	—	399

TABLE B–6 Execution Time in Cycles for the Test Programs.

	Tint	Tlong	Float	Array	Pointer	Arith	Sieve
American Automation	2,109	16,9741	266,489[1]	2,711	1,423	7,686	9,828
Archimedes	1,325	4,488	57,732	774	1,141	2,929	4,699
Avocet	1,168	1,802	150,211	1,393	944	721	4,351
BSO/Tasking	2,268	7,238	—	736	335	6,519	5,142
Franklin	842	5,630	76,179	700	1,392	726	4,699
Intermetrics	1,127	7,082	67,517	948	468	518	4,691
MCC	7,073	—	—	8,445	—	—	34,468
Dunfield	2,318[1]	—	—	3,874[1]	1,850[1]	—	26,101[1]

Notes:
1. Tested using large model.

documentation, ease of installation, resource requirements of the host system, availability for different hosts, embedded system support features, portability, cost of obtaining support, availability of updates, the vendor's commitment to the product, and so on. I will attempt to address these issues throughout this article. The fact that a problem or feature of a specific compiler is discussed, does not necessarily mean that it is missing in the other compilers. I hope this section will give you a feeling for each company and a few more questions to ask when inquiring about various compiler features.

AMERICAN AUTOMATION

The documentation offered by American Automation comes in an $8^{1}/_{2}$-by-11-inch binder. It is nicely laid out, and in general, quite readable. It contains a complete C reference manual that is also useful when problems arise in respect to what C is supposed to be. It would be nice if the index had a few more entries.

Floating-point numbers are supported only by what is called the standard model (roughly equivalent to large model). The compiler also supports assembly through #asm and #endasm preprocessor options. This compiler had the slowest compilation time of the bunch and, like several others, requires an intermediate assembly phase.

I had problems running several of the benchmark programs on this compiler. I later discovered a mistake of my own, but in the meantime I had uploaded the software to their bulletin-board system and had them compile and run it for me. The bulletin-board system is a fairly easy way to get support, and I noticed that application notes and software updates for registered users were available there as well.

ARCHIMEDES

Some time after I started seriously looking at 8051 compilers, I attended two presentations; one by Intermetrics and the other by IAR, the Swedish parent of Archimedes. I proposed the idea of a review to representatives of both companies just after that meeting. The IAR representatives gave me a copy of their compiler on the spot. They seemed very eager to have their product reviewed, as well I can understand.

The Archimedes compiler is the only one that comes with perfect-bound manuals—all the others come in binder form. The cost of this type of binding becomes competitive with three-ring binders only in quantities above 5,000 pieces. It would appear they plan to sell a lot of compilers. These manuals have the disadvantage of not laying flat on the table when you try to read and type at the same time.

Their compiler is also the only one that supports bank switching, and since ANSI compatibility has been their goal for years, I suspect that their C-8051 compiler is one of the most compatible in the field, although I didn't run any specific tests for this. (For example, their interrupt procedures can be declared using pragmas.)

Version 4 of the Archimedes compiler supports recursion but not reentrancy Archimedes allows reentrancy in the slower v. 3 package, which is still actively supported. It must be a nuisance for users who need the features of both to support two separate versions of the compiler. They require a rather complex linker command file to be set up before a program will run. The degree of control over placement of code and data is useful on larger projects. Although Archimdedes provides an example command file that will work with slight modification, a completely prefabricated version of the command file would be nice.

This compiler package is complete and reasonably easy to use. The company is committed to maintaining and improving their compiler, and it is well worth the price.

AVOCET

Avocet began making cross assemblers for CP/M systems in 1979. I have always associated their name with the old "put the photocopied papers in the plastic sack with the disk" approach to documentation. Reviewing their C compiler, AvCase, has significantly changed that impression. Their documentation was the smartest looking in the group. It was enclosed in three boxed binders, and neatly indexed with 38 plastic tabs in color-coded groups.

The package contains the compiler, assembler, linker, librarian, a MAKE utility, editor, and integrated-environment shell similar to Borland's Turbo C. The C compiler generates an assembly language file, which is then passed to the assembler. Although the industry trend is toward generating object files directly, the only thing I have against a separate pass for the assembler is the potentially longer compile time. As the Avocet package had one of the fastest compile times, I would say that's not a problem here.

The original version Avocet shipped to me came with an 8086 compiler. I thought the inclusion of the compiler was a nice touch, and they were the only company that did so. I am told that in the two months since then, they have dropped the 8086 product, illustrating how quickly the offerings change.

The only significant disadvantage I noted was the lack of a compiled stack. Overall, the Avocet compiler is a solid product. Another release is planned for July. Reportedly, this version will support a compiled stack and improved code generation. Neither the code quality nor the execution speed of the current version are compelling reasons to buy this compiler, but as good as their documentation is, their next release is worth watching for.

BSO/TASKING

BSO/Tasking appears to have borrowed heavily from the original Intel tools for the 8051. Their assembler is fully compatible with the Intel assembler, including its powerful text substitution macro language. BSO/Tasking supports a compatible PL/M compiler, which can generate code that may be linked with code from the C compiler. Franklin, who also seems to have ties to the Intel tools, mentions a limit of 256 global symbols imposed for "historical reasons," which BSO/Tasking assured me has been fixed in their product.

BSO's documentation is well organized, clear, and easy to read. It is not as extensive as Archimedes' or Avocet's documentation but I had no trouble finding answers to the questions I had.

I noticed two interesting features browsing through the documentation. First, they support several built-in functions to allow access to special 8051 instructions, such as the test and clear bit (JBC) and the decimal adjust (DAA), in a way that is portable, yet generates fast-in-line code. Second, they fold string constants so that identical constants in the same module are automatically allocated the same memory area. In applications that output a great deal of text, arrays of pointers can be set up to your message strings, and the compiler can compress the string constants for you.

The first package I received from them was copy protected with a parallel-port dongle. Fortunately, they have given up the copy protection and now, like the other packages reviewed here, theirs comes without it—an important concession on their part, as I feel copy protection is simply unacceptable in a development package. I still have some code written under CP/M nearly 10 years ago that I have to maintain and modify occasionally. The tools used to generate it now run under a simulator on my PC. Imagine trying to do that with a copy protected package.

With the exceptions of floats, all the features that I consider important in an 8051 C compiler appear to be present. While their code speed is not at the top of the list, it is reasonably good and the code size is very competitive. Library source is included, and with the floating-point support to be released shortly, it will be a good value.

DUNFIELD SHAREWARE

Micro-C is a shareware package written by Dave Dunfield. The distribution disks contain the full source code for the compiler, assembler, libraries, and approximately 20 useful utilities, including a version of MAKE and an 8051 monitor program. This compiler is designed to be retargeted and, in addition to the 8051 compiler, the disk contains code to build compilers for the Z80, 8086, 6809, and 68HC11.

Micro-C is, like the Micro/C-51 compiler from Micro Computer Controls, a compiler for a subset of C. Like MCC's product, Micro-C doesn't support floats, longs, or structures. Unlike MCC's compiler is *does* support pointers to pointers, multidimensional arrays, parametered defines and compound assignment operators (*=, for example). It doesn't generate relocatable code, but has a unique source linker that links assembly source files and then assembles the entire linked package. It seems a bit clumsy, but it works.

Obviously, this package is not in the same league as most of the others. If you have a serious project, you should look at one of the professional packages. However, this package might do the job for small projects, and if you are a hacker, it has a lot of code for you to play with.

FRANKLIN

Franklin leads the pack in code generation. Their compiler consistently generated the smallest code, with the exception of Intermetrics' pointer code. In regard to speed, the results were mixed but generally good. It supports floating-point and long numbers in a models. The more recent version of the compiler (v 3.0) supports reentrancy and recursion. The older version doesn't, but it is still supported by Franklin and generates good code. In fact, the older v. 2.5 package at $650 offers the best price and performance tradeoff of all the compilers I tested.

The manual I originally received with the compiler was rather difficult to read. I received a new, significantly improved copy recently, but the documentation doesn't quite meet the standard set by Avocet, Archimedes, or BSO. All the information is there, although it could be easier to find. (Bigger type and bolder headings might do it.) This also seems like a good place to echo a complaint I saw in another review. Why do manuals always come in kits? After unwrapping and inserting the manual pages in the binder, I flipped through it for half an hour trying to figure out where all the plastic dividing tabs were supposed to go. I still haven't figured all of them out. Maybe perfect-bound manuals aren't so bad after all.

With the exception of a few I/O functions, Franklin doesn't provide library source code. Although I've never needed the source code, I feel more comfortable having it available. I'm also not sure what they are protecting by not letting it out. The Free Software Foundation provides a complete GNU 68000 C Compiler in source code free of charge, and people are still buying commercial 68000 C compilers. There must be a lesson somewhere.

I was a bit disappointed to find that this compiler won't generate an assembly language file that can be assembled. It generates a mixed code and C listing that can't be modified and reassembled. If you want to use assembly language, you'll have to separately assemble a routine and manually link it in. I was also struck by their corporate attitude; a distinct bluntness. No compiler vendor wants to teach a customer C—that's the customer's job. Still, they are the only vendor who tells you so in their manual: "we. . . do not teach you how to code in the C programming language. . .that is your area of expertise."

Despite these complaints, it's still one of the best compilers available, and in my opinion, the best choice if you want to work in single-chip mode. On the other hand, if you have the ROM space it might be worth sacrificing some of it and using a compiler with better documentation.

INTERMETRICS

In addition to the presentation by Archimedes, the other source of inspiration for this review was a presentation by Intermetrics. While Archimedes provided me a copy of their compiler immediately, it took nearly two months to get a copy of the Intermetrics compiler. When I received the manual for the compiler, I discovered they had packaged the benchmark programs I requested in the box that appeared to be the distribution disks. Unfortunately, the actual distribution disks had been left out.

As the deadline for the article approached, *Embedded Systems Programming* sent me their set of the disks. Unlike most of the other high-end compilers, it didn't come with an install program, and the INSTALL.DOC file on the first disk was written in French!

(All of the high-end compilers discussed in this article, except those by American Automation and MCC, are created by or in conjunction with foreign companies.) Another search located a file called INSTAL.DOC containing English instructions, and eventually I got the disks installed.

This compiler is, without question, the most difficult to use. It runs several compilation phases, assembly, and link, all controlled by an executive and a macro language. An enormous number of options are supported, and the macro that controls the compile-assemble-link process can be changed by the user. In the course of compiling the benchmark programs, I discovered that turning on the debug option caused the assembly pass to generate branch-out-of-range errors with source code that compiled successfully without the option. I also found an option (nowiden) whose apparent function is to cause chars to be promoted to ints when calling a function by placing a random number in the upper 8 bits. The documentation says this option is provided to allow the generation of smaller, faster code. It also silently breaks otherwise good code.

Intermetrics' compiler, like several of the others, is available in a standard version and enhanced version. The enhanced version adds several memory models in addition to the complete library source.

The manuals were quite extensive, showing the maturity of a product that has been around for many years. Although the print quality was poor, the layout was easy to follow and the myriad options for the compiler, linker, and assembler were thoroughly documented. The documentation included a section on how to modify the C executive-command file and a section covering the complete machine library—including the utility function calls that are put in by the code generator to do things that are too big to be put inline.

Aside from the bureaucratic bumbling and the bugs in operation, the only feature I found lacking in this compiler was support for the compiled stack. This product has several memory-allocation schemes that allow for static allocation and overlay of local variables. By properly partitioning the functions in different files, the effect of a compiled stack could be obtained, although with a lot of effort on your part.

To be fair, most of the benchmarks used in this review were derived from code that Intermetrics supplied. The product manager at Intermetrics provided me with pages of benchmark test results and analysis that convinced me they are seriously committed to developing a superior product. If they could make their user interface easier to use and bulletproof without losing the complier's other features they would be well on their way to that goal. As it stands, they have a powerful collection of tools but their compiler is also harder to use and has more quirks than those produced by the other companies.

MICRO COMPUTER CONTROLS

Micro Computer Controls is really on the edge of being able to call their product a C compiler. It doesn't support floating-point numbers, long numbers, structures, or multi-dimensional arrays. Function prototyping is not available. Defines are not allowed to have parameters. MICRO/C-51 is comparable to a slightly enhanced version of Small C ported to run on the 8051. It generates source files that must be assembled with Intel's or MCC's 8051 assembler. The assembler is a separate product. If you don't already have an assembler, the complete package is $525.

The documentation is typewritten and very hard on the eye. It does explain how to use the package and its options and also includes several fairly extensive example programs.

Despite the features the product lacks, it does have the essentials. As a historical note, Micro Computer Controls was the first company to try to market an 8051 C compiler in the U.S., but with Franklin's low-end package at $650, and Dunfield's for $40, it seems as if $525 is a bit too expensive for this package. At a lower price, however, it could be useful to someone who didn't have a lot of time-critical code.

C HOW THEY RUN

What do I think are the best products? Archimedes and Franklin are at the top of my list. Franklin for tight code and ease of use and Archimedes for its full complement of features and good documentation. I would be willing to spend my money on either of these products. Right behind them are BSO/Tasking and Avocet. The product from BSO is reasonably fast without generating excessive code. Avocet's compiler has superb documentation.

As far as I'm concerned, the big winner is C. Choose any of these compilers today and you'll be better off than writing in assembly. If you're still doing all your 8051 programming in assembly language, it is time for a change.

Appendix C

Hardware Information

CONVENTIONAL 8051-FAMILY BOARDS

The material that follows is a copy of an article by Richard Poindexter from the March-April issue of *Midnight Engineering* (pp. 16–27) which he has graciously permitted to be included here. It provides a good survey of the broadly advertised 8051-family boards aside from the occasional "evaluation boards" provided briefly by the *chip* manufacturers to promote new versions. The latter may be nice, but they have no commitment to long-term availability and are not likely to be supported as a board. Of course, the commercial designs may change also, but the information here should provide some leads for sources.

Several months ago a friend and I were discussing a business he is involved in. He revealed to me that with the ability to measure and control a few critical physical parameters, he could become a dominate force in his industry. When he explained the details of the physical parameters, I could not believe how basic and simple they were from an engineering point of view. I volunteered to do the research and development of just such a device.

My first order of business was to develop an embedded controller based data acquisition and control system. Since this is a project to be done "after hours" of my regular full time job, I chose not to design my own embedded controller board, but instead to use an "off the shelf" board. My special added value would be to develop the software and sensors to complete the package.

My survey of the embedded controller boards on the market found the most common ones utilized the Intel 8051 family. This article is based on the results of that survey.

Figure C–1   Iota

IOTA SYSTEMS

The Iota Systems EC-32 is an excellent example of a very well designed embedded controller board. The EC-32 has lots of features included on board as well as a very flexible setup configuration. Also, the way the EC-32 and accessories are priced, you pay for only the features and capabilities that you need.

I had the opportunity to review all the products on the Iota Systems price sheet. It includes the EC-32, BASIC-52 (in a 27C256 EPROM) and MONITOR-52 (in a 27C256 EPROM). Each has its own manual. Also available is a PC interface cable, adapter and wall plug-in power module. Also available from Iota are two very useful and low-cost PC disks. One is a Terminal Emulator program. The other disk is an assortment of User Supported 8051 programs. Included are an assembler, disassembler, editor and BASIC compiler. There also is a prototype breadboard called the EC-BB which is a 3.5″ × 5.3″ board with plated through holes on one-tenth of an inch centers and a 50-pin header.

The EC-32 measures 3.5″ × 5.3″ and has a total of 13 ICs. They include the 80C32, 74HC373, 74HC138, MAX232, EPROM, RAM, EEPROM as well as other logic and control gates. The 80C32 has less power consumption than the 8031 and 256 bytes internal RAM compared to 128 bytes in the 8031. Note the EC-32 has three memory sites which can include any combination of 8K or 32 K and EPROM, RAM or EEPROM. The RAM is battery backed up with a 3 volt Lithium battery. On board the EC-32 is a prototyping space measuring 1.2 × 3.5 inches. The prototyping area is adjacent to the 50 pin expansion header. The EC-32 also has an on board +5 Volts DC voltage regulator.

The package also included a great Application Note on how to interface a LCD display module the EC-32. The note includes a schematic, assembly language and BASIC-52 routines, as well as information about where to get the LCD modules. According to literature received from Iota, they offer several different Application Notes. Contact Iota for more information.

The EC-32 User's Manual is an 8 1/2 by 11 inch, 50 page, spiral bound manual. The manual is very well laid out and covers all aspects of the EC-32. The manual included an introduction, quick start directions, hardware hookup and development of projects. Also included are examples of interface projects, schematics, and a resource directory.

Using the Terminal Emulator program and the directions in the Quick Start Chapter, I was able to get set up quickly. Since BASIC-52 and MONITOR-52 comes on EPROM rather than in the internal ROM of the microprocessor, they are easy to install and to use as well as remove. After the PC interface cable is hooked up as well as plugging in the wall power supply, the EC-32 came right up. The MONITOR-52 and BASIC-52 EPROMs are very powerful, yet relatively inexpensive. I was able to pick functions from each and execute them with complete satisfaction.

Figure C–2   Blue Earth

BLUE EARTH

For persons who have little or no hardware background and/or have no desire to spend a minute longer than necessary to include an embedded controller in their next project, the Blue Earth MICRO-440 is just the ticket. Everything about the Blue Earth product is class. From full color brochures and literature to bound manuals, the MICRO-440 is great.

The package that I reviewed was the Macro Assembler Package. The package includes the MICRO-440, the Applications Module, the User Manual, MCS-51 Reference Manual, PC interface cable, wall plug-in power supply, pocket reference guide, A51 macro assembler, OHS-51 file converter, MonScope 51 symbolic hardware debugger and MS-DOS Software Manual.

The three manuals are hefty, bound, paperback "data book" style books. They remind me a lot of the classic Don Lancaster cookbooks. Tom Bachmann and the folks at Blue Earth have logically laid out the documentation such that one can "get a little or a lot" of information to assist with the use of the MICRO-440.

The MICRO-440 itself is a very well thought out design. This controller comes fully fleshed out. Several controllers reviewed here are priced with no ram, monitors or basic. If you want those options, you pay as you go. Not with the MICRO-440. It comes loaded with Basic and Monitor-51 in the ROM of the 83C51FB microcontroller. Also, included is 32K of onboard RAM. The MICRO-440 includes an 8 channel, 8 bit A/D converter, the ADC0848. Another feature is the real time clock/calendar chip, the RTC72423 and a 3.6 volt lithium battery.

Although these are impressive features, they are not the main attraction. The big hitter is the size. The MICRO-440 is implemented with surface mount technology and multilayer printed circuit board. The outside dimensions of the MICRO-440, including the ABS plastic case are 2.95″ × 2.44″ × .75″. If your application is especially limited for space, the MICRO-440 deserves serious consideration.

The combination of simple packaging and super manuals makes getting started easy. Following directions from the User Manual in a section called Quick Start I unpacked, set up, and loaded software on a PC in less than 10 minutes. Blue Earth offers a separate hardware device which is the same size and shape as the MICRO-440, called the Applications Module. It plugs into one end of the MICRO-440. It has several built-in functions including a seven-segment LED display, sound transducer, two IRL530 MOSFET transistors and a 10K potentiometer. Within 25 minutes of unpacking, I was using some canned BASIC routines to exercise the MICRO-440 and the Applications Module. If the need for other devices are required for use with the MICRO-440, a project board with matching plastic case is available for prototyping.

To summarize the MICRO-440; it has the most thorough and complete documentation of all boards reviewed. Also, it is the physically smallest of all the boards reviewed. However, if the MICRO-440 is used in applications requiring expansion, the required DB-25 connectors make for a long flat controller—difficult to squeeze into the typical space available. Overall, the MICRO-440 is a top notch product.

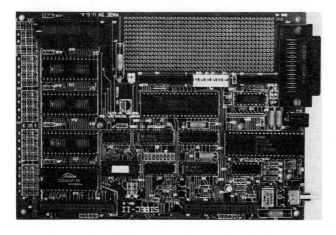

Figure C–3   Binary Technology

BINARY TECHNOLOGY

The SIBEC-II is a moderately sized embedded controller board that offers the 8052-AH BASIC V1.1 CPU with floating point BASIC interpreter as standard equipment. The EPROM Programmer found on the SIBEC-II is the only one of all the 8051 embedded controller boards reviewed which offers EPROM programming.

The products reviewed from Binary Technology included the SIBEC-II single board BASIC controller, SIBEC-F3 Monitor/Debugger EPROM and the BTK52.

The SIBEC-II Single Board BASIC Controller measures 5 3/8″ × 7 3/4″. The board comes standard with 11 ICs, with sites for 3 more memory devices as well as the Zero Insertion Force EPROM programmer socket. The devices employed include the 8052-AH, 8255, 6264, 2 - 74LS373, 74LS245 and some miscellaneous control and gate logic. A feature important for product development is prototyping space. The SIBEC-II includes five square inches of space with holes on 0.1 inch centers. Additionally there are four unused chip selects available for quick prototype work.

The EPROM programmer is capable of programming both 2764 and 27128 devices, and at 12.5 or 21 volts. A special programming power supply is required if the programming capability is utilized. Also not found on other boards are two iSBX expansion connectors. There is space for two single width iSBX cards or one double width card.

Other hardware products available from Binary Technology but not reviewed here are the SIBEC-II/8031 and the SIBEC-IIPB. The SIBEC-II/8031 Single Board Controller has the same features as the SIBEC-II except no EPROM programming hardware. Included with the SIBEC-II is a 8031 microcontroller rather than the 8052-AH chip. Also included is the SIBEC-F2 monitor/debugger. The SIBEC-11PB Production Board has the same features as the SIBEC-II except no EPROM programming hardware, iSBX connectors or reset button. It also comes equipped with an 8031 rather than the 8052-AH microcontroller.

The SIBEC-II Hardware Documentation is a 14 page spiral bound manual. It includes circuit description, jumper position directions, memory map, 8255 instructions, PROM programming and iSBX Expansion Connector. The SIBEC-F3 Monitor/Debugger Program Documentation is a 16 page spiral bound manual which includes a summary description, command summary, BASIC extensions, monitor resource utilization, user accessible routines and installation.

The SIBEC-II is worth your consideration particularly if you require BASIC, EPROM Programming and iSBX expansion plug in connectors.

ALLEN SYSTEMS

A unique, high performance board called the DP-31/535 is offered by Allen Systems. A dual processor board which has sockets for either a standard 40 pin 8051 family device or a 80535, but not at the same time. Also featured are up to 120K of EPROM or RAM, an 8 channel 8 bit A/D and up to six 8 bit parallel ports.

The products I reviewed included the DP-31/535 embedded controller board as well as

Figure C–4   Allen Systems

the SM-31/535 System Monitor EPROM and DB-51D/8051 Enhanced System Monitor EPROM.

Although I am reviewing only the DP-31/535 it is worth mention of other ALLEN SYSTEMS products. They include the CP-31, which is an 80C31 single board computer. Also available is the CP-526, which is an 80C535 single board computer. It is especially unique because of its 82526 CAN (Controller Area Network) chip, which was originally introduced for automotive high-speed serial interfaces.

The DP-31/535 measures 6.5″ × 4.125″, has 12 ICs including the 80535 or 8031 (or another 40 pin 8051 family member), 8256 peripheral expansion chip, 74LS373 address latch, 74LS138 address decoder, EPROM, RAM and other various logic and glue chips. Also included are sockets for two more RAM or EPROM devices.

The 80535 includes common features from the 8032 and 8052, but in addition includes an 8 channel 8 bit A/D converter, 2 more 8 bit parallel ports, twelve interrupts with four priority levels and a 16 bit watchdog timer.

The DP-31/535 User's Manual includes a block diagram, memory type/location/map jumpers, connector pinout, parts list and schematic. The SM-31/535 User's Manual includes resource utilization, command summary and description, download format, system configuration and interrupts and program listing. The DB-51 User's Manual is the same as SM-31/535 with the exception of no program listing.

The PC-40 is a unique and versatile prototyping card available from Allen Systems not directly related to the DP-31/535. It is a small card measuring 4.5 ″ × 2.85″ with a 27 × 20 matrix of holes on .1 inch centers adjacent to a 2 × 20 header and a 40 pin DIP socket. The bottom of the PC-40 is plugged into the target board and the IC is plugged into the PC-40.

In my testing of the DP-31/535, I had the opportunity to use the board with the 8031 installed (remember only the 8031 or the 80535 can be plugged in at a time) and the SM-31/535 system monitor program. The System Monitor software provides the capability to examine or change internal or external memory, both data or program. Also available is the ability to examine or change registers, move a block of memory, single step or execute a program and download Intel hex object code. The DP-31/535 is a neat idea in terms of its dual processor scheme. The high performance of the 80535 with its expanded I/O and built-in A/D converter make the DP-31/535 a real hard charger.

TRI-L DATA SYSTEMS

From Tri-L Data Systems comes the GPC11 Single Board Computer. The GPC11 sports a 16 or 20 MHz 80C535, 5 × 4 keypad encoder, 8 channel 8 bit A/D converter, five 8 bit parallel ports and up to 64K EPROM and 32K RAM. Software support consists of a library of over 40 utilities written in C.

The products reviewed included the GPC11 Single Board Computer, GPC Monitor, GPC11 C Utilities, MONTERM Terminal Emulation program. Measuring 3″ × 5″ the GPC11 has 11 ICs. Included are the 80C535, 88C681 DUART, 74HC373 address latch, 74HC155 address decoder, 74C923 keypad encoder, 27C256 EPROM and RAM as well as communications transceivers.

The GPC11 Technical Manual is a 72 page spiral bound manual. It includes a circuitry description, jumper settings, connector definitions, schematic, parts list and memory map. Two-thirds of the manual is devoted to a discussion of the C source programs. For each routine a summary, description, return value and an example of usage is given.

The GPC Monitor Board is an important accessory for use with the GPC11. It is a small plug-in board to be used as a development tool. To use the GPC Monitor Board the EPROM, RAM and a jumper on the GPC11 must be removed, then the Monitor Board is plugged into Expansion Bus Connector P01. Programs can be downloaded into memory from the host PC. Using the functions of Monitor-51 in the Monitor Board, the programs can be debugged and cleaned up. After program development is complete, the Monitor Board is unplugged and the EPROM, RAM and jumper are returned to the GPC11. The Monitor Boards 26 page User's Manual gives Monitor-51 commands and instructions as well as MONTERM Terminal Interface information.

The C routines shipped with each GPC11 include C source code to initialize serial ports, transmit and receive at serial ports, an interrupt handler for serial ports, read keypad, set and read real time clock, initialize A/D and A/D conversion. Other C code includes routines to do binary to ASCII decimal conversion and back, print message to serial ports, start DUART timers/counters, bcd to integer and integer to character.

When I set up the GPC11 and the GPC Monitor Board as described above and then cabled up the serial interface to a PC, the system worked just fine. The Monitor-51 commands worked just as if it were installed physically on the GPC11 itself. Furthermore, the MONTERM terminal emulation program for the PC performed satisfactorily

As a part of the standalone C routines, included on the distribution diskette are separate subdirectories depending on whether the routines will be used with an Archimedes or Franklin C compilers. The startup C code which the compiler needs before it can compile and link programs is included along with a compiler specific command file to link the main routine to produce executable, downloadable or absolute code is included. The software development utilities, along with the C routines themselves, can make a valuable contribution to your 8051 product development library.

The combination of the GPC11, the GPC Monitor Board and the library of C source code makes a super package. If your application requires a large amount of I/O lines and you could use the jump start at software development provided by the C source code, then you should consider the GPC11.

Figure C–5   Modular Micro Controls

MODULAR MICRO CONTROLS

The TR-537 is one of the top performance single board computers from Modular Micro Controls. The company name is an exact statement of what the TR-537 and the entire system of products is all about: MODULARITY. To an engineer who has been schooled in the virtues of modular design, using the Modular Micro Controls is a treat.

The TR-537 measures 3.93″ × 8.2″ and has 10 ICs. They include the 80C537, 74HC573 address latch, 74HC245 buffer, 26C256 RAM, 27C256 EPROM as well as other logic and memory map control. The Siemens 80C537 includes five 8-bit ports, 4 timer/counters with capture and compare using up to 21 pulse width modulation/high speed outputs, 12 channel 8-bit A/D converter with programmable voltage reference and 14 interrupts with 4 levels.

The Technical Reference Manual includes an overview, circuit description, memory map setup, I/O port description, serial port information, expansion header configuration, schematics and specifications.

The products received from Modular Micro Controls for review were numerous, but they were only a small sample of the huge system of products offered by Micro Modular Controls. The design goal was to provide an engineer with a system of products that can be used for prototyping new designs very rapidly. The system is composed of four types of boards: CL, TR, MC and PP.

The CL or Cluster are inexpensive, small single board computers for simple appli-

Comparison Chart

Company Board #	Allen Systems DP-31/535	Binary Technology SIBEC-II	Blue Earth MICRO-440	Cottage Resources Control-R II	Iota Systems EC-32	Micromint RTC31	Modular Micro TR-537	Tri-L Data CPC11
Price	$250	$228	$199	$64.95	$100	$119	$189	$275
Processor	8031/80535	8052-AH	83C51FB	8031	80C32	8031	80C537	80C535
Clock (Mhz)	12.00	11.0592	12.00	11.0592	11.0592	11.0592	11.0592	12.00
EPROM size	8, 16 or 32K	8K	None	8 or 16K	8 or 32K	8 or 32K	8, 32 or 64K	32 or 64K
RAM size	8 or 32K	8K	32K	8K	8 or 32K	8 or 32K	8 or 32K	8 or 32K
EEPROM size	None	None	None	None	8 or 32K	None	32K	None
RS-232?	Yes, 2 CH	Yes	Yes, 2 CH.	Yes	Yes	Yes	Yes, 2 CH.	Yes, 2 CH.
Speed			to 19.2K	4800	9600	to 19.2K	375K	to 19.2K
Current (Typ)	200 mA	350 mA	70 mA		35 mA	30 mA	200 mA	66 mA
Dimensions	4 1/8" × 6 1/2"	5 3/8" × 7 3/4"	2.95" × 2.44"	2.75" × 4.00"	3.5" × 5.3"	3.5" × 3.5"	3.93" × 8.2"	3" × 5"
Exp. Headers?	Yes	Yes	Yes	Yes	Yes	Yes	Yes	Yes
# pins	70	72	50	74	50	52	190	78
BASIC?	Yes	Yes	Yes	No	Yes	Yes	Yes	No
Opt. Cost	$40	No charge	No charge		$10	$25	Call	
Monitor?	Yes	Yes	Yes	No	Yes	Yes	Yes	Yes
Opt. Cost	$35, $100	$75	No charge		$40	$50	Call	$150
A/D?	Yes	No	Yes	No	No	Yes	Yes	Yes
Opt. Cost	No Charge		No Charge				Call	No Charge
Clock/Calendar	No	Yes	Yes	No	No	No	Yes	Yes
Opt. Cost		$39	No Charge				Call	$35
Prototype Area?	No	Yes	No	No	Yes	No	No	No
Bare Board?	Yes	Yes	No	Yes	No	No	Yes	No
Cost	$75	$45		$25			Call	

cations. The TR series are general purpose single board computers based on the EURO-CARD footprint. The MC or Modular Controller series is just that: a Modular set of electronic circuit building blocks with at least one memory or peripheral function per MC module. Finally the PP or Port Proto series are port adapters designed to allow the user to modify any I/O port.

There are CL and TR boards available based on the following 80C31, 87C751, 87C752, 80C552, 80C851 and of course the 80C537. The MC series is a very dynamic and growing family of modules which includes 24 modules at this time. Among these are the MC-PIC, -PROTO, -HOST, -PORT, -MEMORY, -MIC, -A/D, -D/A, -ARCNET, -UART and -MODEM. All MC modules include software support modules in both assembly and C language along with the hardware.

My experience with the TR-537 clearly demonstrates the modular power of the Modular Micro Controls system. I began by simply plugging an MC-PORT module into the TR-537, then plugging in a PP-LED, PP-DIP to the MC-PORT and then plugging in a PP-485 module directly into a port header on the TR-537. I had set up and used the RS-485 communications with Port Status LEDs and switch support in literally minutes.

An engineer with a moderate set of modules could be ready to plug together boards, link together software modules and have a full blown system within hours. Compare this to standard methods of taking a single board computer and "rolling your own" for peripherals and I/O; and you can see this system is an engineer's dream!

COTTAGE RESOURCES

The Control-R II from Cottage Resources Corporation has the most simplistic design and is the least expensive of all the 8051 embedded controller boards reviewed. The Control-R II is actually a second generation product. Yes, there is a Control-R board. Although it is not reviewed here it is worth mention. At only $39.95 for a populated board (minus MAX232) or $15 for the blank board, it is the low-cost entry level board. It consists of only 4 chips; they are the 8031, 74LS373, 2764 (programmed with Softest program) and MAX232 on a single sided PC board.

The main disadvantage of the Control-R is the lack of data storage memory beyond 128 bytes in the internal RAM in the 8031. The Control-R II solves that problem with the addition of an 8K Static RAM (6264) to the Control-R design. Also added is a 74LS02 for address decoding.

The design approach Cottage Resources has taken for the Control-R II is to use a textbook 8031 circuit design, select the memory device types, physical locations and memory map configuration. The benefit of hard wiring these choices is that it reduces the number of jumpers headers, gate logic and pc board pads and traces which are needed for a more flexible configuration.

The documentation consists of 14 pages devoted to theory of operation, expansion header pinout, schematic, Softest ROM discussion, and an example assembly language program for use with an analog to digital converter and serial interface.

Included with the Control-R II is a 2764 EPROM programmed with the Softest program. The Softest program consists of special assembly language routines to check out the Control-R II for proper operation and to demonstrate some of its functions. One function reads the status of an input port and copies it back to an output port. Another outputs a software controlled oscillation. Also included are interactive functions. They include the ability to display the contents of the 8031's internal 128 bytes of RAM, display the first 256 bytes of program memory and then display the input status of Port I.

As you have seen, the Control-R II has the lowest cost of all the embedded controller boards reviewed. In some budgets or with particular applications which are very price sensitive the cost of the embedded controller is the number one priority. If the lack of flexibility with memory sizes, types, layout, map is less of a concern than the cost, then certainly the Control-R II is worth your careful review.

MICROMINT

The Micromint RTC31/52 is a very busy, dense design. The RTC comes with either an 8031 or 80C52, depending on whether or not you want built in BASIC-52. The package I reviewed was the RTC 52 EDS (Economical Development System). The package includes the RTC 80C52 BASIC Controller board (with 8K RAM installed), the BASIC-52 manual and the BASIKIT which is a software development package.

The RTC measures 3.5" × 3.5" and consists of nine IC's (when fully loaded with EPROM and RAM). Among these are the 8031 (or 80C52), 74HCT245 address/data buffer, 74HCT373 address latch and MAX232 serial interface chip, EPROM, Static RAM as well as miscellaneous glue logic chips.

One advantage of the RTC31/52 is that it includes DDT-51 headers. The DDT-51 (Dynamic Debugging Tool) was one of the most popular projects Steve Ciarcia ever did for Byte. For about $100 an engineer had the ability to have a low-cost 8051 family development system that allows one to stop at breakpoints, single-step through instructions, examine and/or change register and memory locations. Not exactly the same as a $1000 plus in-circuit emulator, but the functions are very helpful with 8051 based designs.

The documentation is moderately helpful. Most of the manual is devoted to jumper setting directions for memory device type/size selections, memory map layout and device location selection. Other attention is given to schematics, expansion headers, bill of materials and 8051 family background.

A strong advantage for the RTC31/52 is the RTC family of boards available from Micromint. Other than the RTC31/52 there are several RTC expansion bus compatible First, there are three other processor boards available; RTC-HC11 (68HC11 processor) RTC180 (64180 processor), RTCV25 (V25 processor). Next, there are severa input/output expansion boards. They are the RTCIO, which is a Multifunction Expansion Board and the RTCBUFIO, which is a Buffered Expansion Board. Also there are the RTC-OPTO, which is an optically isolated I/O Expansion Board; the RTC-SIR which is a Serial and Infrared I/O Expansion Board; and the RTC-LCD, which is a display, keyboard and X-10 I/O Expansion Board.

The RTC31/52 is a very strong embedded controller board. It is important especially if your application requires functions such as are available in the wide variety of boards in the RTC family.

CONCLUSION

This review is by no means exhaustive. I would have no problem recommending every one of the 8051 embedded controller boards reviewed. As with any product, the potential customer must define what is important and make the selection based on which controller board has the features needed. If you are still debating between two or more product offerings, most companies will sell you just the manual. Reviewing the manuals can help resolve remaining doubt about which board best fits your needs. When corresponding with any of these vendors of 8051 embedded controller boards, be sure to mention that you heard of them through *Midnight Engineering* magazine!

Figure C–6 Bitbus Cards

BITBUS CARDS

Intel Products

Intel has two cards with BITBUS support (there are also 2 cards that tie in to either a SBX port on a Multibus host or a PC card that installs in a PC host). The first application card is the iRCB 44/10, which has the core electronics along with a set of 3 8-bit parallel ports. The card includes sockets for repeater drivers when using long runs. It can be configured to run the tasks out of RAM (downloaded over the bus) or to start up with user tasks in EPROM. The card has a single SBX connector which can hold any of a number of small piggy-back cards such as a serial I/O, additional parallel I/O, or A-D or D-A cards. There may even be IEE 488 bus interface modules or floppy disk interfaces.

The second card is the iRCB 44/20 which replaces the parallel ports of the 44/10 with analog I/O functions. It includes a 12-bit A-D with an 8/16 channel multiplexer as well as 2 12-bit D-A output channels. The same SBX capability is there as well.

The two other Intel cards are the SBX 344 and the PCX 344. The former has hardware to interface to an SBX port and the latter is equipped to go into a PC. Incidentally, the former card can attach to a 44/10 or 44/20 card to allow a gateway between two separate BITBUS networks.

Gespac Products

Gespac supplies a system they call *FILBUS*, which is a hardened version of BITBUS. The product line includes several different cards for analog and digital I/O oriented toward industrial control racks. The boards have an enclosure, a set of indicator LEDs, and a built-in switching power supply. There is a lot of isolation employed to protect the system from damage. The system is promoted with on-board firmware preconfigured for typical applications. For example, the digital card can be set up as a self-contained scheduler with individual I/O lines going high and low at set times. Analog cards can put out programmed wave shapes. In the various cases, key parameters can be sent down from the host and the node can continue to run unattended. It is said, however, that the boards can be run as conventional BITBUS cards as well.

DEVELOPMENT RESOURCES

In-Circuit Emulators

While not discussed in detail, most of the industrial development efforts can be greatly helped by emulators. They are hardware/software systems which allow development to take place without the burden of the processor supporting a mon itor and serial I/O for the developer. Instead, the emulator plugs into the socket where the processor usually resides.

Monitor Programs

Much development work can be done using the target controller as the debugging tool. By installing a firmware (EPROM) program which starts on power-up, it is possible to communicate with, say, a PC which supports the compiler or assembler being used for code development. A small program on the PC can interface to the target system over a serial link and the monitor program allows the user to download new code to RAM on the target (with the PSEN and RD ored together). Then the monitor program allows the user code to be run, perhaps with single-stepping or break points. The various areas of memory can be examined and changed when the monitor program is running. The main drawbacks over the emulators is the necessity to include serial ports, special memory jumpering to allow writing to code space for downloading, and the need to locate the user code away from the 0000 start-up location. It is not difficult, but the *final* burned-in code will have to be located differently and will require a few changes in the interrupt vectors. It is definitely a more economical way to undertake program development.

RTOS Development Resources

BITBUS monitor.   If you are running development for DCX and BITBUS, this monitor is quite useful. It is a program that runs on a host computer (PC under DOS or Multibus under RMX) that enables the development of tasks on remote boards via messages sent over the BITBUS network. In other words, you can download and run tasks on the 44/10 and 44/20 cards by typing commands on a host computer. The interaction is possible because remote boards have a communication task (the RAC task, described later) present in the firmware with the operating system firmware. The monitor software allows all the RAC task activities including the sending of data to the remote RAM. With the proper jumpering choice the RAM and code space can overlay each other so the monitor can load data into the program space. It takes a large number of messages to load a long program with 13-byte messages, but the transmission rate is high enough to be acceptable—perhaps a minute to load a 4K program.

In addition to downloading, it is possible to start the tasks (c r e a t e t a s k), delete tasks, send and receive messages, and examine any memory locations *while the tasks are running*. That is because the RAC task is high priority and preempts the other tasks. Thus, it is possible to examine variables while the application is running as opposed to having to stop with breakpoints. The Bitbus monitor is an essential development tool for multitasking applications with DCX.

MON51 for DCE.   A very similar product to the BITBUS monitor, this runs with DCE-COMM to allow the same sort of functionality over a serial network based on RS-422 that BITBUS obtains with RS-485 and SDLC. It is in preliminary stages at this point, but bears watching for distributed systems.

UBI (Universal Bitbus Interface).   While the BITBUS monitor allows a user to examine and control remote nodes by typing commands, the UBI allows the same sort of commands to be executed out of programs running on the host computer. In other words, whatever you can do manually, you can include in a program on the host. Thus, the remote nodes can be addressed, examined, have programs downloaded from the local mass storage devices all under program control. User-friendly interfaces are possible and systems can be developed where the remote nodes are an integral part of data collection or control jobs. Many of the BITBUS applications actually use nothing but the RAC task on the remote nodes and treat the cards as smart, serial-linked I/O systems.

RAC (Remote Access and Control) Task.   Intel's 44/10 and 44/20 boards include this task in firmware. It is just another task in the overall scheme of DCX, but it is the key to debugging and cold start-up before tasks are put in EPROM. The commands defined in the messages sent to this task allow the reading and writing of the memory spaces as well as the creation of other tasks and the examination of the status of the node. It can also take messages to be sent to the I/O ports so it can allow the I/O to be used directly from a separate host.

Appendix D

Vendors—Addresses, Phones, and Products

It would be foolish to think that this list will remain complete for 6 months, let alone for the (hopefully long) life of this book (one company was dropped during the writing and one was added). However, from experience it would seem that *some* leads are better than none, and one good lead can open up to a host of others. The following information is what I have available in early 1992.

MAGAZINES

Embedded Systems Programming
  Miller Freeman Publications
  600 Harrison Street
  San Francisco, CA 94107
  (415) 905–2200

Midnight Engineering
  111 East Drake Road
  Suite 7041
  Ft. Collins, CO 80525
  (303) 225–1410
  FAX (303) 225–1075

COMPILERS AND ASSEMBLERS

Intel Corporation
  3065 Bowers Ave.
  Santa Clara, CA 95051
  (408) 986–8086
  FAX (408) 727–2620

Franklin Software, Inc.
  888 Saratoga Ave. #2
  San Jose, CA 95129
  (408) 296–8051

Archimedes Software
  2159 Union St.
  San Francisco, CA 94123
  (415) 567–4010

BSO/Tasking
  128 Technology Center
  PO Box 9164
  Waltham, MA 02254
  (617) 894–7800

Avocet Systems Inc.
  120 Union St.
  Rockport, ME 04856
  (800) 448–8500

American Automation Inc.
  14281 Chambers Rd.
  Tustin, CA 92680
  (714) 731–1661

REAL-TIME OPERATING SYSTEMS

(DCX)
  Intel Corporation
  3065 Bowers Ave.
  Santa Clara, CA 95051
  (408) 986–8086
  FAX (408) 727–2620

United States Software Corp.
  14215 N.W. Science Park Drive
  Portland, OR 97229
  (503) 641–8446
  FAX (503) 644–2413

(DCE)
  Iota Systems, Inc.
  PO Box 8987
  Incline Village, NV 89450–8987
  (702) 831–6302
  FAX (702) 831–4629

Byte-BOS Integrated Systems
  PO Box 3067
  Del Mar, CA 92014
  (619) 755–8837

CMX Company
  19 Indian Head Heights
  Framingham, MA 01701
  (508) 872–7675

(RTX and RTXtiny)
  Franklin Software, Inc.
  888 Saratoga Ave. #2
  San Jose, CA 95129
  (408) 296–8051

(RTXC)
  A T Barrett and Associates
  11501 Chimney Rock, Suite R
  Houston, TX 77035
  (713) 728–9688

CHIPS

Intel Corporation
  3065 Bowers Ave.
  Santa Clara, CA 95051
  (408) 986–8086
  FAX (408) 727–2620

Siemens Components, Inc.
  2191 Laurelwood Road
  Santa Clara, CA 95054
  (408) 980–4500
  FAX (408) 980–4596

Signetics Company
  811 E. Arques Avenue
  PO Box 3409
  Sunnyvale, CA 94088–3409
  (408) 991–2000

Oki Semiconductor Inc.
785 N. Mary Ave.
Sunnyvale, CA 94086
(408) 720–1900

Dallas Semiconductor
4350 Beltwood Pkwy.
Dallas, TX 75224
(214) 450–0400

Fujitsu Microelectronics
3545 N. First St.
San Jose, CA 95134
(800) 642–7616

COMMERCIAL BOARDS

Iota Systems, Inc.
PO Box 8987
Incline Village, NV 89450–8987
(702) 831–6302
FAX (702) 831–4629

Micromint
4 Park Street
Vernon, CT 06066
(203) 871–6170

Modular Micro Controls
109 South Water St.
Northfield, MN 55057
(507) 645–8315

Binary Technology
PO Box 67
Meriden, NH 03770
(603) 469–3232

Allen Systems
2646 Brandon Rd.
Columbus, OH 43221
(614) 488–7122

Blue Earth Research
410 Belle Ave.
Mankato, MN 56001
(507) 387–4001

Cottage Resources
1405 Stevenson Dr.
Suite 3–672
Springfield, IL 62703
(217) 529–7679

BITBUS BOARDS

Intel Corporation
  3065 Bowers Ave.
  Santa Clara, CA 95051
  (408) 986–8086
  FAX (408) 727–2620
Gespac
  50 West Hoover Ave.
  Mesa, AZ 85210
  (602) 962–5559
  FAX (602) 962–5750

DEVELOPMENT HARDWARE

Intel Corporation
  3065 Bowers Ave.
  Santa Clara, CA 95051
  (408) 986–8086
  FAX (408) 727–2620

Appendix E

RTOS System Calls

	DCX(DCE)[1]	(RTX)tiny[2]	USX[3]	CMX[4]	Byte-BOS[5]	RTXC[6]
TASK MANAGEMENT (sometimes a distinction exists between delete and suspend)	createtask deletetask (suspendtask) (resumetask)	createtask deletask	runtsk klltsk pritsk slttsk scdtsk	tcre trmv tpri tend twatm twake twakf	createtask runtask sleep pause deactivatetask endtask halttask resumetask denyreschedule forcereschedule wakeup wakeupall wakeuphipri wakeuppri	execute terminate
EVENTS, FLAGS, AND SEMAPHORES (Some are simple shared flags while others are counters. There is a significant variation here.)		sendsignal clearsignal wait	setevt clrevt incevt decevt chkevt wteset wteclr setgrp	ecre esig ewatm eread fgsig fgwatm fgread	enableevent ackevent sleepforevent sleepforallevents eventoccurred waitforevent waitforallevents disableevent eventenabled postevent posteventall posteventhipri posteventpri	signal signalm wait waitm waitt pend pendm

(*Continued*)

	DCX(DCE)[1]	(RTX)tiny[2]	USX[3]	CMX[4]	Byte-BOS[5]	RTXC[6]
RESOURCES (a resource is managed as a simple flag)			chkres getres reqres relres	rscre rsget rsrsv rsrel	initresource captureresource captureresourcenb releaseresource releaseresourcehipri	lock lockw lockt unlock
MESSAGES (all quite similar except for timeout and handshake features)	sendmessage (returnmessage) wait	(sendmessage) wait	sndmsg rcvmsg chkmsg	mssend mssdst mssenw msswst msget mswatm mswakc	putmessage putmessagenb enablemessage messageoccurred ackmessage getmessage disablemessage messageenabled sleepformessage waitformessage	send sendw sendt receive receivew receivet defmbxsema
MEMORY BLOCKS (a few are on-chip but most are off-chip)	allocate (getmem) deallocate (releasemem)	(createpool) (getblock) (freeblock)	reqmem chkmem relmem	bfcre bfget bfrel		alloc allocw alloct free

(Continued)

						RTXC
TIME (some of these calls involve periodic timers while others provide simple delay or timeout)	wait setinterval	wait (setslice)	dlytsk wketsk setclk getclk period	tpcre tpstt tpstp tpread	delay wait timeoutoccurred disabletimeout enablesampling enabletimeout acktimeout updatetimers	alloctimer alloctimeout starttimer stoptimer restarttimer freetimer freetimeout
QUEUES, LISTS UART, MISC.	getfunctionids			lcre lrst latt latb lrft lrfb	(24 uart calls will not be listed here)	enqueue enqueuew enqueuet dequeue dequeuev dequeuet defqsema purgequeue
INTERRUPTS (a marked difference in power here)	disableinterrupt enableinterrupt	isrsendsignal (attachinterrupt) (detachinterrupt) (disableisr) (enableisr) (isrsendmessage) (isrrecvmessage)	usxcmd_c	intin intex	enterisr exitisr	isrc

[1] DCX calls are all preceded by RQ as in rqwait. Calls in parentheses are only in DCE.
[2] RTX calls are preceded by OS_ and words are separated by underline as well. Calls in parentheses are not in RTXtiny.
[3] USX calls are all 6 characters long but have no special prefix.
[4] CMX calls all begin with cx but have no separation characters.
[5] Byte-BOS calls all begin with bb_ and words are separated by underlines.
[6] RTXC calls all have KS_ prefixes and words are separated by underlines.

Index